Global maths - Mathematics without limits

Contents

Introduction

Decolonising UK mathematics

Theoretical Foundations

The Philosophy and History of Mathematics

Equity and Inclusion in Mathematics Education

Curriculum Content and Context - Diversify Mathematical Knowledge

   Identifying Underrepresented Contributions in Mathematics

   Integrating Diverse Mathematical Practices and Theories

Contextualising Mathematics

   Importance of Contextualising Mathematics:

   Incorporating Real-World Problems Relevant to Diverse Communities

   Examples of Contextualised Mathematics Teaching

Pedagogical Approaches Understanding Culturally Responsive Pedagogy

   Principles of Culturally Responsive Pedagogy

   Strategies for Engaging Students from Diverse Backgrounds

   Teacher Training and Professional Development

Collaborative and Inclusive Classroom Practices

   Practicalities of Promoting Collaborative Learning Environments

   Inclusive Teaching Strategies to Support All Learners

   Addressing Bias and Stereotypes in the Mathematics Classroom

Implementing Change - Curriculum Design and Assessment

   Practicalities of Designing a Decolonised Maths Curriculum Framework

   Inclusive Assessment Practices and Tools

   Continuous Improvement Through Feedback and Reflection

Policy and Institutional Support - the Impact of Educational Policies

   Practicalities of Designing a Decolonised Maths Curriculum Framework

   Examples of Successful Policy Implementation in the Mathematics Classroom

Global Mathematics – Knowledge and ideas

   Global mathematics in West Africa:

Example Lesson outline ideas and Activities:

Activities and Projects Promoting Inclusive Mathematics Learning:

Signposts for diversity in mathematics

Annotated Bibliography: Decolonising Mathematics Education in the UK

Global Maths – Mathematics without limits

Introduction

Decolonising UK mathematics

Decolonisation in education refers to the process of critically examining and transforming the curriculum, teaching practices, and educational policies to dismantle the legacy of colonialism. It aims to create an inclusive and equitable educational environment that reflects diverse perspectives and knowledge systems. Decolonisation is significant because it seeks to rectify historical injustices and provide a more holistic and accurate representation of global knowledge.

Significance:

- Decolonisation promotes educational equity by valuing the contributions of all cultures and ensuring that no group is marginalised.
- It encourages critical thinking by challenging dominant narratives and fostering a deeper understanding of global knowledge systems.
- An inclusive curriculum that reflects students' diverse backgrounds can enhance engagement and motivation, making learning more relevant and meaningful.

Historical Context of Colonialism in Education

The historical context of colonialism in education is marked by the imposition of Western knowledge systems and cultural values on colonised peoples. During the colonial era, education systems were designed to serve the interests of the colonisers, often disregarding, and devaluing indigenous knowledge and practices.

Impact on Education:

- Curriculum content predominantly reflects European perspectives and achievements, often to the exclusion of other cultures.
- Western knowledge is often positioned as superior, creating hierarchies that marginalise non-Western ways of knowing.
- Indigenous languages, histories, and traditions have been systematically suppressed or erased in educational contexts.

Relevance to the UK Maths Curriculum

The relevance of decolonisation to the UK maths curriculum lies in addressing the implicit biases and assumptions that shape the teaching and learning of mathematics. Mathematics is often perceived as a culturally neutral and objective discipline, but this perception overlooks the rich mathematical contributions of non-European cultures and the context in which mathematical knowledge is applied.

Key Points for Mathematics Teachers:

- Teachers should acknowledge and incorporate the mathematical contributions of various cultures, such as the advancements made by ancient civilisations in Africa, Asia, and the

Middle East (Joseph, 2011). For instance, the concept of zero and algebra have significant origins in Indian and Islamic mathematics.

- Educators need to critically evaluate the curriculum to ensure it does not solely highlight European mathematicians and theories. Highlighting the global development of mathematical ideas can provide a more comprehensive understanding of the subject (Pais, 2017). And it can support mastery of concepts and theories

- Adopting culturally responsive teaching methods can make mathematics more accessible and relevant to all students. This includes using real-world examples that resonate with students' diverse backgrounds and experiences (Gay, 2010). This can affect academic performance positively.

Implementing Decolonisation in the Maths Curriculum:

Understanding and implementing decolonisation in the UK maths curriculum is essential for creating an equitable and inclusive educational environment. By acknowledging the historical context of colonialism in education and actively working to incorporate diverse perspectives and contributions, mathematics teachers can help dismantle the legacy of colonialism and promote a more holistic and accurate understanding of mathematics.

Starting points are to examine the arithmetic curriculum on a regular basis to spot and correct any Eurocentric biases. Incorporate works by non-European mathematicians and investigate ideas in mathematics that originated outside of Europe (Bidwell, 1993). Give educators continual training on global perspectives on mathematics history and culturally appropriate teaching methods. To create a more inclusive learning environment, teachers can also encourage students to research and present on mathematical ideas and figures from other cultural contexts.

References:

- Bidwell, J. K. (1993). "Humanise Your Classroom with the History of Mathematics." The Mathematics Teacher, 86(6), 461-464.

- Gay, G. (2010). Culturally Responsive Teaching: Theory, Research, and Practice. Teachers College Press.

- Joseph, G. G. (2011). The Crest of the Peacock: Non-European Roots of Mathematics. Princeton University Press.

- Pais, A. (2017). "Critique and Politics of Mathematics Education." Educational Studies in Mathematics, 96(1), 137-153.

Theoretical Foundations

The Philosophy and History of Mathematics

Understanding the Philosophy and History of Mathematics

The Philosophy of Mathematics explores the nature, origins, and implications of mathematical concepts. It addresses questions about the foundations of mathematics, its truth, and its role in human understanding. For teachers, understanding the philosophy of mathematics can provide deeper insights into how mathematical knowledge is constructed and interpreted.

The History of Mathematics traces the development of mathematical ideas and practices across different cultures and time periods. This historical perspective reveals the diverse contributions that have shaped mathematics as we know it today.

Ancient Origins and Evolution of Mathematics

Mathematics has ancient and diverse origins, evolving through the contributions of many cultures and civilisations over millennia. Understanding these origins challenges the notion that mathematics is a purely European invention.

1. African Mathematics:
    - The ancient Egyptians developed sophisticated techniques for arithmetic, geometry, and algebra to solve practical problems related to agriculture, architecture, and astronomy (Gillings, 1972).
    - The Rhind Mathematical Papyrus, dating back to around 1650 BCE, is one of the oldest known mathematical texts. It provides insight into the rich and foundational mathematical findings Africans.

2. West Asia - Babylonian Mathematics:
    - Babylonian mathematicians made significant advances in algebra and developed a base-60 number system, which is still used today in measuring time and angles (Friberg, 2007).
    - The Plimpton 322 tablet, dating to around 1800 BCE, contains an advanced understanding of Pythagorean triples.

3. Indian Mathematics:
    - Ancient Indian mathematicians like Aryabhata and Brahmagupta made groundbreaking contributions to algebra, arithmetic, and trigonometry. The concept of zero as a number was first recorded in India (Joseph, 2011).
    - The work of Indian scholars was transmitted to the Islamic world and later influenced European mathematics.

4. Chinese Mathematics:

- Chinese mathematicians developed early techniques in algebra, geometry, and number theory. The "Nine Chapters on the Mathematical Art," dating back to around 100 CE, is a foundational text (Chemla & Guo, 2004).
- They also used a decimal place value system long before it appeared in Europe.

5. Islamic Mathematics:
   - During the Golden Age of Islam, scholars like Al-Khwarismi and Omar Khayyam made significant advancements in algebra, trigonometry, and calculus (Berggren, 1986).
   - Islamic mathematicians preserved and expanded upon Greek, Indian, and Persian mathematical knowledge, which was later transmitted to Europe during the Renaissance.

Challenging Eurocentric Narratives

The traditional Eurocentric narrative in mathematical history often overlooks or marginalises the contributions of non-European cultures. This narrative tends to present mathematics as a linear progression of ideas centred on European achievements, particularly during the Renaissance and Enlightenment periods. Consider that theories now named after Greek or Italian mathematicians for example had previously been formulated and recorded by ancient African civilisations centuries before. However, supremacist and colonialist mindsets ignored, poached, renamed or destroyed many discoveries in Africa.

Relevance to the UK Maths Curriculum:

- Broadening Perspectives: Incorporating the global history of mathematics into the curriculum provides a more accurate and comprehensive understanding of the subject. This helps students appreciate the diverse cultural contributions to mathematics and challenges the misconception that it is a purely Western discipline.
- Enhancing Engagement: Acknowledging the contributions of various cultures can make mathematics more relatable and engaging for students from diverse backgrounds, fostering a sense of inclusion and respect for their heritage (D'Ambrosio, 2001).
- Promoting Critical Thinking: Encouraging students to critically examine the history and philosophy of mathematics helps them understand the subject as a dynamic and evolving field, rather than a static body of knowledge (Pais, 2017).

Implementing a Diverse Mathematical History in the Classroom

1. Curriculum Content: Integrate historical examples and problems from different cultures into the curriculum. Highlight the achievements of mathematicians from various backgrounds.
2. Teacher Training: Provide professional development opportunities for teachers to learn about the global history of mathematics and how to incorporate it into their teaching.
3. Resources and Materials: Develop and use teaching materials that reflect the diverse origins of mathematical concepts. Use texts and resources that present a balanced view of mathematical history.

Conclusion

Understanding the philosophy and history of mathematics is crucial for providing a comprehensive and inclusive mathematics education. By exploring the true and ancient origins and evolution of mathematics and challenging Eurocentric narratives, mathematics teachers can enrich their students' learning experiences and promote a more equitable and accurate understanding of the subject.

References:

- Berggren, J. L. (1986). Episodes in the Mathematics of Medieval Islam. Springer.
- Chemla, K., & Guo, S. (2004). Les Neuf Chapitres: Le Classique Mathématique de la Chine Ancienne et ses Commentaires. Dunod.
- D'Ambrosio, U. (2001). Ethnomathematics: Link Between Traditions and Modernity. Sense Publishers.
- Friberg, J. (2007). A Remarkable Collection of Babylonian Mathematical Texts. Springer.
- Gillings, R. J. (1972). Mathematics in the Time of the Pharaohs. MIT Press.
- Joseph, G. G. (2011). The Crest of the Peacock: Non-European Roots of Mathematics. Princeton University Press.
- Pais, A. (2017). "Critique and Politics of Mathematics Education." Educational Studies in Mathematics, 96(1), 137-153.

Equity and Inclusion in Mathematics Education

Educational equity in mathematics education refers to the principle of fairness, ensuring that all students have access to the necessary resources, opportunities, and support to succeed, regardless of their background or personal circumstances. It involves creating an inclusive environment where every student can engage with and understand mathematical concepts, thereby achieving their full potential.

Importance of Educational Equity:

- Fairness and Justice: Educational equity addresses systemic inequalities, providing all students with a fair chance to succeed (Gutiérres, 2008). It ensures that no student is disadvantaged due to factors beyond their control.

- Maximising Potential: Ensuring equitable access to quality education allows all students to reach their full potential, benefiting both individuals and society (Boaler, 2016). Diverse talents and perspectives are fostered, enriching the educational experience for everyone.

- Social Cohesion: Promoting equity in education helps to build a more inclusive and cohesive society by bridging gaps between different social and cultural groups (Nieto, 2010). It fosters mutual respect and understanding among students from diverse backgrounds.

Barriers to Inclusion in Current Maths Curricula

Several barriers to inclusion exist within the current mathematics curricula, which need to be addressed to promote equity:

1. Cultural Biases:

    - Mathematics curricula often reflect a Eurocentric and supremacist perspective, highlighting contributions from European mathematicians while neglecting or poaching those learned from other cultures (Pais, 2017). This can alienate students from non-European backgrounds, making it harder for them to see themselves as capable mathematicians.

2. Stereotypes and Expectations:

    - Implicit biases and stereotypes about who is 'good' at maths can negatively impact students' self-perception and performance (Archer et al., 2013). For example, the stereotype that boys are naturally better at maths than girls can discourage female students from pursuing mathematical excellence.

3. Resource Disparities:

    - Schools in underprivileged areas often lack access to high-quality inclusive resources and experienced teachers, which can impede students' learning and engagement with mathematics (Wilkins, 2010). This resource gap perpetuates educational inequalities.

4. Pedagogical Practices:

    - Traditional teaching methods that focus on rote learning and individual performance can disadvantage students who benefit from collaborative and context-based learning approaches (Boaler, 2016). Such methods may not cater to diverse learning styles and needs.

The Role of Identity and Culture in Learning Mathematics

Identity and culture play a crucial role in students' learning experiences and achievements in mathematics. Understanding this relationship is essential for creating an inclusive and equitable maths curriculum.

1. **Cultural Relevance:**
   - Integrating culturally relevant examples and contexts into the mathematics curriculum can make the subject more engaging and meaningful for students (Gay, 2010). For instance, using real-world problems that resonate with students' lives and communities can enhance their interest and understanding.

2. **Mathematical Identity:**
   - Students develop a sense of mathematical identity through their interactions with teachers, peers, and the curriculum. Positive reinforcement and representation of diverse mathematicians can help all students see themselves as capable of excelling in mathematics (Nasir, 2002). Encouraging diverse mathematical role models and histories can help students relate to and value the subject.

3. **Language and Communication:**
   - Language plays a significant role in how students understand and engage with mathematical concepts. Providing support for students who speak English as an additional language and using inclusive language practices can help ensure all students can fully participate in maths lessons (Setati, 2005).

Relevance to the UK Maths Curriculum:

By incorporating a wider range of cultural contributions and perspectives into the maths curriculum, teachers can help students appreciate the global nature of mathematics. This can challenge the Eurocentric bias and make the subject more inclusive (Joseph, 2011).

Adopting teaching practices that recognise and value students' cultural backgrounds can enhance their engagement and achievement. Culturally responsive teaching methods, such as collaborative learning and real-world problem-solving, can cater to diverse learning needs (Gay, 2010).

Ongoing professional development for teachers on issues of equity and inclusion is crucial. This can include training on recognising and addressing implicit biases, as well as strategies for creating an inclusive classroom environment (Nieto, 2010).

Conclusion

Understanding and promoting equity and inclusion in mathematics education is vital for ensuring all students in England have the opportunity to succeed. By addressing cultural biases, challenging stereotypes, and recognising the role of identity and culture in learning, teachers and school leaders can create a more inclusive and equitable maths curriculum. This will not only improve educational outcomes but also contribute to a more just and cohesive society.

References:

- Archer, L., DeWitt, J., Osborne, J., Dillon, J., Wong, B., & Willis, B. (2013). "Not girly, not sexy, not glamorous": Primary school girls' and parents' constructions of science aspirations. *Pedagogy, Culture & Society*, 21(1), 171-194.

- Boaler, J. (2016). Mathematical Mindsets: Unleashing Students' Potential through Creative Mathematics, Inspiring Messages and Innovative Teaching. Jossey-Bass.

- Gay, G. (2010). Culturally Responsive Teaching: Theory, Research, and Practice. Teachers College Press.

- Gutiérres, R. (2008). A "Gap-Gasing" Fetish in Mathematics Education? Problematising Research on the Achievement Gap. *Journal for Research in Mathematics Education*, 39(4), 357-364.

- Joseph, G. G. (2011). The Crest of the Peacock: Non-European Roots of Mathematics. Princeton University Press.

- Nasir, N. S. (2002). Identity, goals, and learning: Mathematics in cultural practice. *Mathematical Thinking and Learning*, 4(2-3), 213-247.

- Nieto, S. (2010). The Light in Their Eyes: Creating Multicultural Learning Communities. Teachers College Press.

- Pais, A. (2017). "Critique and Politics of Mathematics Education." *Educational Studies in Mathematics*, 96(1), 137-153.

- Setati, M. (2005). Teaching Mathematics in a Primary Multilingual Classroom. *Journal for Research in Mathematics Education*, 36(5), 447-466.

- Wilkins, C. (2010). Social class and the classroom: Key issues in teaching and learning. *Primary Mathematics*, 14(1), 5-10.

Curriculum Content and Context - Diversify Mathematical Knowledge

Diversifying mathematical knowledge involves recognising and incorporating the contributions of underrepresented groups and cultures into the mathematics curriculum. This approach enriches students' understanding of mathematics by highlighting its global development and application.

Identifying Underrepresented Contributions in Mathematics

The history of mathematics is replete with contributions from various cultures that are often overlooked in traditional curricula. Highlighting these contributions can challenge the Eurocentric narrative and provide a more accurate representation of the field. How many of the following are regularly referred to in Mathematics classrooms?

1. **African Mathematics:**

    - The Ishango Bone, dating back to around 20,000 BCE, found in the Democratic Republic of Congo, is considered one of the earliest tools used for counting and arithmetic (Saslavsky, 1999).
    - Ancient Egyptian mathematics included sophisticated techniques for geometry and algebra, which were used in constructing pyramids and other architectural marvels (Gillings, 1972). Imhotep's work in constructing pyramids involved understanding their geometric properties. Various African Papyri held by Europeans hold vast foundational African mathematical knowledge for example from the 'The Rhind Mathematical Papyrus'.
    - The ancient Egyptians had a unique numerical system and mathematical operations. They used symbols to represent numbers and performed calculations through combining symbols for addition, removing symbols for subtraction, repeating symbols for multiplication, and subtracting repeatedly for division.
    - In ancient Egypt, African people used their own set of symbols to represent numbers. These symbols were called hieroglyphs.

2. **Indian Mathematics:**

    - Indian mathematicians like Aryabhata and Brahmagupta made significant contributions to algebra and trigonometry. The concept of zero as a number and the decimal place value system were developed in India (Joseph, 2011).
    - The work of Indian mathematicians was later transmitted to the Islamic world and Europe, influencing the development of modern mathematics.
    - Vedic Mathematics refers to a collection of ancient techniques and principles described in ancient Indian scriptures, known as the Vedas. These methods are efficient for solving complex arithmetic calculations and algebraic problems.
    - Integration into Curriculum: Introduce Vedic techniques for quick mental arithmetic and algebra, demonstrating their applications alongside traditional methods (Joseph, 2011).

3. **Chinese Mathematics:**
    - The "Nine Chapters on the Mathematical Art" is an ancient Chinese text that covers a wide range of mathematical topics, including arithmetic, geometry, and solving linear equations (Chemla & Guo, 2004).
    - Chinese mathematicians also developed early methods for solving quadratic equations and working with fractions.
    - Ancient Chinese mathematicians used rod numerals and the abacus for calculations. These tools facilitated complex arithmetic operations and influenced the development of computational methods.
    - Integration into Curriculum: Incorporate lessons on the history and use of the abacus, allowing students to explore alternative computation methods and understand their historical significance (Chemla & Guo, 2004).

4. **Islamic Mathematics:**
    - Scholars like Omar Khayyam made groundbreaking advancements in algebra, trigonometry, and calculus during the Islamic Golden Age (Berggren, 1986).
    - Islamic mathematicians preserved and expanded upon the knowledge from Greek, Indian, and Persian sources, which later influenced European mathematics during the Renaissance.
    - During the Islamic Golden Age, mathematicians like Al-Khwarismi made the introduction of systematic solving of linear and quadratic equations.
    - Integration of the curriculum: Teach the history of algebra, highlighting Al-Khwarismi's work and its influence on modern mathematics. Include problem-solving techniques from his texts in the curriculum (Berggren, 1986).

Integrating Diverse Mathematical Practices and Theories

Integrating diverse mathematical practices and theories into the curriculum involves more than just adding historical facts; it requires a shift in how mathematics is taught and perceived.

1. Curriculum Content:
    - Include contributions from various cultures in textbooks and lesson plans. For example, when teaching algebra, discuss Al-Khwarizmi's work and its impact on the field.
    - Highlight mathematical concepts developed independently by different cultures, such as the use of zero in India and the Mayan civilisation.
2. Pedagogical Approaches:

- Use culturally responsive teaching methods that connect mathematical concepts to students' cultural backgrounds and everyday experiences (Gay, 2010).
- Employ project-based learning that encourages students to explore mathematical problems from a global perspective, considering how different cultures have approached similar challenges.

3. Professional Development:
    - Provide teachers with training on the history and contributions of non-Western mathematical systems. This can include workshops, seminars, and access to resources that offer a global perspective on mathematics (Nieto, 2010).
    - Encourage collaboration and sharing of best practices among teachers to develop more inclusive teaching strategies.

Relevance to the UK Maths Curriculum

Integrating diverse mathematical knowledge into the UK maths curriculum can:

- Enhance Engagement: Providing a broader perspective can make mathematics more engaging and relevant for students from diverse backgrounds.
- Foster Inclusivity: An inclusive curriculum acknowledges and values the contributions of all cultures, promoting a sense of belonging and respect among students.
- Develop Critical Thinking: Encouraging students to explore and compare different mathematical approaches fosters critical thinking and a deeper understanding of the subject.

Conclusion

Diversifying mathematical knowledge is crucial for creating a more inclusive and accurate representation of the field. By identifying and integrating underrepresented contributions and using culturally responsive teaching methods, mathematics teachers in England can enrich their students' learning experiences and promote a more equitable education.

References:

- Berggren, J. L. (1986). Episodes in the Mathematics of Medieval Islam. Springer.
- Chemla, K., & Guo, S. (2004). Les Neuf Chapitres: Le Classique Mathématique de la Chine Ancienne et ses Commentaires. Dunod.
- Gay, G. (2010). Culturally Responsive Teaching: Theory, Research, and Practice. Teachers College Press.
- Gillings, R. J. (1972). Mathematics in the Time of the Pharaohs. MIT Press.
- Joseph, G. G. (2011). The Crest of the Peacock: Non-European Roots of Mathematics. Princeton University Press.
- Nieto, S. (2010). The Light in Their Eyes: Creating Multicultural Learning Communities. Teachers College Press.

- Saslavsky, C. (1999). Africa Counts: Number and Pattern in African Cultures. Chicago Review Press.

Contextualising Mathematics

Contextualising mathematics involves teaching mathematical concepts through real-world problems and scenarios that are relevant to students' lives. This approach makes mathematics more engaging and meaningful, helping students see the practical applications of what they are learning.

Importance of Contextualising Mathematics:

- Enhancing Relevance and Engagement: By linking mathematical concepts to real-world situations, students can better understand the importance and applicability of mathematics in everyday life (Boaler, 2016).
- Promoting Critical Thinking: Contextualising problems encourages students to think critically and creatively about how to apply mathematical principles to solve complex issues (Schoenfeld, 2014).
- Supporting Diverse Learners: Using examples from various cultures and communities can make mathematics more inclusive, reflecting the diverse backgrounds of students (Gay, 2010).

Incorporating Real-World Problems Relevant to Diverse Communities

To make mathematics instruction relevant and inclusive, teachers should incorporate real-world problems that reflect the diverse communities of the UK and the broader global context. Here are some strategies and examples:

1. Local Contexts:
   - Urban Planning: Use problems related to urban planning and architecture, such as designing a park or planning efficient public transportation routes. These problems can involve geometry, algebra, and data analysis.
   - Local Economy: Explore mathematical concepts through the lens of local businesses and economy, such as calculating costs, profits, and budgeting for a local shop or market. This can involve arithmetic, percentages, and financial mathematics (Wake, 2011).

2. Global Contexts:
   - Climate Change: Address global issues like climate change by analysing data on carbon emissions, temperature changes, and the effects of various mitigation strategies. This can involve statistics, calculus, and mathematical modelling (Hine, 2018).
   - Global Health: Use problems related to global health, such as modelling the spread of diseases or analysing the effectiveness of vaccination programs. This can involve statistics, probability, and exponential functions (Bloom, 2017).
3. Cultural Relevance:
   - Ethnomathematics: Incorporate mathematical practices and concepts from different cultures, such as African fractals, Islamic geometric patterns, and Mayan numeration systems. This approach not only enriches students' understanding but also validates the mathematical contributions of diverse cultures (D'Ambrosio, 2001).
   - Culturally Relevant Examples: Use examples that reflect the cultural backgrounds of the students, such as calculating the ingredients for a traditional dish or planning a cultural festival. This makes learning more personal and engaging (Ladson-Billings, 1995).

Examples of Contextualised Mathematics Teaching

Project-Based Learning:

A project on designing an eco-friendly school building. Students can use geometry to design the layout, algebra to calculate materials needed, and data analysis to evaluate energy efficiency. This project connects mathematics to sustainability and environmental science (Boaler, 2016).

Data Analysis Projects:

Analysing local traffic patterns to propose improvements. Students can collect data, create graphs, and use statistics to identify problem areas and suggest solutions. This real-world application enhances their understanding of data handling and statistical analysis (Schoenfeld, 2014).

Mathematics and Art:

Exploring symmetry and geometry through art. Students can study Islamic geometric patterns or African fractals, learning about the mathematical principles behind these designs and creating their own artwork. This approach integrates mathematics with cultural studies and art (D'Ambrosio, 2001).

Sports Analytics:

Analyse sports statistics to teach concepts like averages, percentages, and probability. Students can calculate batting averages in cricket or baseball, shooting percentages in basketball, or win-loss ratios in various sports. This approach demonstrates the practical application of mathematics in sports analysis and decision-making.

Consumer Mathematics:

Explore concepts of budgeting, saving, and spending through real-world scenarios related to consumer finance. Students can calculate discounts, interest rates, and monthly payments when buying a car, taking out a loan, or managing a budget. This approach helps students develop financial literacy skills essential for everyday life.

Environmental Science:

Use mathematical modelling to understand environmental phenomena like population growth, deforestation, and climate change. Students can analyse data on wildlife populations, carbon emissions, or temperature trends to explore exponential growth, linear regression, and other mathematical concepts. This approach integrates mathematics with environmental science and sustainability.

Social Justice Mathematics:

Investigate social justice issues using mathematical analysis and data interpretation. Students can examine disparities in income distribution, access to healthcare, or educational opportunities using statistical techniques. This approach fosters critical thinking about social issues and encourages students to advocate for change using mathematical evidence.

Engineering and Design:

Apply mathematical principles to engineering and design challenges. Students can design structures, bridges, or transportation systems using geometry, trigonometry, and calculus. They can calculate forces, angles, and dimensions to ensure the safety and efficiency of their designs. This approach connects mathematics to real-world engineering applications.

Health and Nutrition:

Explore mathematical concepts through the lens of health and nutrition. Students can analyse nutritional labels, calculate calorie intake, and evaluate dietary choices using arithmetic and proportions. They can also study trends in obesity rates, food consumption, and exercise habits using statistical analysis. This approach promotes health literacy and informed decision-making.

Technology and Coding:

Integrate mathematics with computer science and coding activities. Students can learn about algorithms, patterns, and logic through programming languages like Python or Scratch. They can create simulations, games, or data visualisations that demonstrate mathematical concepts and problem-solving strategies. This approach develops computational thinking skills alongside mathematical reasoning.

Cultural Mathematics:

Explore mathematical practices and concepts from different cultures around the world. Students can study mathematical patterns in art, music, or architecture from diverse cultural traditions. They can investigate topics like symmetry in Islamic art, fractals in African design, or geometric constructions in indigenous cultures. This approach celebrates cultural diversity while deepening students' understanding of mathematical principles.

Relevance to the UK Maths Curriculum

The UK maths curriculum emphasises problem-solving, reasoning, and the application of mathematical skills in real-world contexts. Contextualising mathematics aligns with these goals by making learning more relevant and engaging for students.

Implementation Strategies:

- Curriculum Design: Incorporate real-world problems and projects into the curriculum, ensuring they are relevant to students' lives and diverse backgrounds.

- Teacher Training: Provide professional development for teachers on how to effectively contextualise mathematics and use culturally responsive teaching methods (Gay, 2010).

- Collaborative Learning: Encourage group work and collaborative problem-solving, allowing students to share their diverse perspectives and approaches to mathematical problems.

Conclusion

Contextualising mathematics is essential for making the subject relevant, engaging, and inclusive for students in the UK. By incorporating real-world problems that reflect local and global contexts, teachers can enhance students' understanding and appreciation of mathematics. This approach supports the goals of the UK maths curriculum and promotes educational equity and excellence.

References:

- Boaler, J. (2016). Mathematical Mindsets: Unleashing Students' Potential through Creative Mathematics, Inspiring Messages and Innovative Teaching. Jossey-Bass.

- Bloom, B. S. (2017). Global Health and the Role of Mathematics. Journal of Global Health, 7(1), 010201.

- D'Ambrosio, U. (2001). Ethnomathematics: Link Between Traditions and Modernity. Sense Publishers.

- Gay, G. (2010). Culturally Responsive Teaching: Theory, Research, and Practice. Teachers College Press.

- Hine, G. (2018). The Role of Mathematics in Addressing Climate Change. Environmental Education Research, 24(4), 523-536.

- Ladson-Billings, G. (1995). Toward a Theory of Culturally Relevant Pedagogy. American Educational Research Journal, 32(3), 465-491.
- Schoenfeld, A. H. (2014). Mathematical Problem Solving. Academic Press.
- Wake, G. (2011). Modelling School Mathematics: The Case of Financial Mathematics. Educational Studies in Mathematics, 76(3), 285-305.

Pedagogical Approaches Understanding Culturally Responsive Pedagogy

Culturally responsive pedagogy is an approach to teaching that acknowledges and values the cultural backgrounds, experiences, and perspectives of all students. In mathematics education, it involves:

1. **Diverse Cultural Contributions**:
    - The UK maths curriculum should reflect the diverse cultural contributions to mathematics, including examples and problems that resonate with students from different backgrounds (D'Ambrosio, 2001).

2. **Inclusive Learning Environments**:
    - Culturally responsive pedagogy fosters a supportive and inclusive learning environment where all students feel valued and empowered to participate in mathematics learning (Nasir, 2002).

3. **Equitable Access to Education**:
    - By addressing cultural biases and incorporating diverse teaching methods, culturally responsive pedagogy helps ensure equitable access to mathematics education for all students, regardless of their cultural background or identity (Ladson-Billings, 1995).

4. **Promoting Excellence and Achievement**:
    - Culturally responsive teaching promotes academic achievement by engaging students in meaningful learning experiences that connect mathematics to their cultural identities and experiences (Gay, 2010).

Incorporating diverse cultural references, examples, and teaching methods to make the subject more relevant and accessible to students from diverse backgrounds can improve the engagement, performance and outcomes of learners.

Principles of Culturally Responsive Pedagogy

1. **Cultural Awareness**:
    - Teachers should be aware of their own cultural biases and assumptions, as well as those of their students. Understanding and respecting the cultural backgrounds of students is essential for building meaningful connections in the classroom (Gay, 2010).

2. **Inclusive Curriculum**:
    - The curriculum should reflect the diverse cultural contributions to mathematics and include examples and problems that resonate with students from different backgrounds. This helps students see themselves reflected in the subject matter and fosters a sense of belonging (D'Ambrosio, 2001).

3. **Relational Teaching:**
   - Building positive relationships with students based on mutual respect and understanding is crucial for creating a supportive and inclusive learning environment. Teachers should actively seek to connect with students and show an interest in their cultural identities and experiences (Ladson-Billings, 1995).

4. **Student-Centred Learning:**
   - Culturally responsive teaching values student voice and agency, allowing students to contribute their perspectives and experiences to the learning process. Teachers should provide opportunities for students to share their cultural knowledge and connect it to mathematical concepts (Gay, 2010).

Strategies for Engaging Students from Diverse Backgrounds

1. **Incorporate Culturally Relevant Examples:**
   - Use examples, problems, and activities that relate to students' cultural backgrounds and experiences. For example, when teaching geometry, explore patterns in Islamic art or African designs (Nasir, 2002).

2. **Promote Collaborative Learning:**
   - Encourage group work and collaborative problem-solving activities that allow students to learn from each other's perspectives and approaches. This promotes a sense of community and collective responsibility for learning (Gay, 2010).

3. **Use Multimodal Instruction:**
   - Employ a variety of teaching methods and resources, including visual aids, hands-on activities, and multimedia materials, to accommodate diverse learning styles and preferences (Gay, 2010).

4. **Provide Culturally Responsive Feedback:**
   - Offer feedback that recognises and values students' cultural knowledge and contributions. Acknowledge students' diverse perspectives and encourage them to draw connections between their cultural experiences and mathematical concepts (Nieto, 2010).

Teacher Training and Professional Development

1. **Cultural Competence Training:**
   - Teachers need training on cultural competence, including awareness of their own cultural biases and strategies for creating inclusive learning environments (Gay, 2010).

2. **Curriculum Development**:
   - Professional development should focus on developing culturally responsive curriculum materials and lesson plans that integrate diverse cultural perspectives and examples (Nieto, 2010).

3. **Pedagogical Strategies**:
   - Teachers should learn effective pedagogical strategies for engaging students from diverse backgrounds, including collaborative learning, differentiated instruction, and culturally relevant teaching methods (Ladson-Billings, 1995).

4. **Reflective Practice**:
   - Encourage teachers to engage in reflective practice, regularly reflecting on their teaching practices and the impact on students from diverse backgrounds. Provide opportunities for teachers to share best practices and learn from each other (Gay, 2010).

Conclusion

Culturally responsive pedagogy is essential for creating inclusive, equitable, and high-quality mathematics education in the UK. By incorporating diverse cultural perspectives, examples, and teaching methods, mathematics teachers can create learning environments that empower all students to succeed.

References:

- D'Ambrosio, U. (2001). Ethnomathematics: Link Between Traditions and Modernity. Sense Publishers.

- Gay, G. (2010). Culturally Responsive Teaching: Theory, Research, and Practice. Teachers College Press.

- Ladson-Billings, G. (1995). Toward a Theory of Culturally Relevant Pedagogy. American Educational Research Journal, 32(3), 465-491.

- Nasir, N. S. (2002). Identity, goals, and learning: Mathematics in cultural practice. Mathematical Thinking and Learning, 4(2-3), 213-247.

- Nieto, S. (2010). The Light in Their Eyes: Creating Multicultural Learning Communities. Teachers College Press.

Collaborative and Inclusive Classroom Practices

Understanding Collaborative and Inclusive Classroom Practices

Collaborative and inclusive classroom practices in mathematics involve creating learning environments where all students feel valued, supported, and empowered to participate actively in their learning. This approach fosters collaboration, equity, and academic achievement for all students.

Practicalities of Promoting Collaborative Learning Environments

1. Group Work and Peer Collaboration:
   - Organise activities that require students to work in small groups or pairs to solve mathematical problems collaboratively. This encourages peer learning, communication, and teamwork skills (Cobb et al., 2001).

2. Structured Collaboration:
   - Provide clear guidelines and roles for group work to ensure equitable participation and accountability. Rotate roles within groups to distribute leadership and responsibility (Johnson et al., 2014).

3. Teacher Facilitation:
   - Act as a facilitator rather than a lecturer during collaborative activities. Circulate among groups, ask probing questions, and provide support and feedback as needed (Boaler, 2016).

4. Reflection and Evaluation:
   - Encourage students to reflect on their collaborative experiences and evaluate their group dynamics. Use feedback from students to adjust group assignments and improve collaborative practices (Webb, 2009).

Inclusive Teaching Strategies to Support All Learners

1. Differentiated Instruction:
   - Tailor instruction to meet the diverse learning needs and preferences of students. Provide multiple entry points and scaffolding for tasks to ensure all students can access and engage with the material (Tomlinson, 2014).

2. Universal Design for Learning (UDL):
   - Design lessons and materials with flexibility and accessibility in mind, allowing all students to participate and succeed regardless of their abilities or backgrounds. Incorporate multiple modalities and formats for presenting information (Rose & Meyer, 2002).

3. Culturally Responsive Teaching:
   - Integrate examples, problems, and teaching methods that reflect the cultural backgrounds and experiences of students. Validate and value students' cultural identities and contributions to mathematics (Gay, 2010).

4. Explicit Teaching of Metacognitive Strategies:
   - Teach students how to monitor their own learning, set goals, and reflect on their progress. Encourage metacognitive dialogue and self-assessment to help students become more independent learners (Hattie, 2009).

Addressing Bias and Stereotypes in the Mathematics Classroom

1. Critical Examination of Curriculum and Materials:
   - Review curriculum materials and resources for bias and stereotypes, particularly regarding gender, race, and socio-economic status. Choose materials that present diverse representations of mathematicians and mathematical concepts (Martin, 2009).

2. Challenging Stereotypes through Discussion:
   - Engage students in discussions about bias and stereotypes in mathematics and society. Encourage critical thinking and reflection on how stereotypes can influence perceptions and performance in mathematics (Steele, 2010).

3. Counteracting Stereotypes with Positive Representation:
   - Highlight examples of diverse mathematicians and their contributions to the field. Emphasise the achievements of individuals from underrepresented groups to challenge stereotypes and promote inclusivity (Boaler, 2016).

Relevance to the UK Maths Curriculum

Promoting collaborative and inclusive classroom practices aligns with the goals and principles of the UK maths curriculum in several ways:

1. Promoting Equity and Inclusion:
   - Collaborative and inclusive practices support the curriculum's aims of promoting equity, inclusion, and excellence in mathematics education for all students (Department for Education, 2014).

2. Developing Collaborative Skills:
   - Collaborative learning environments help students develop teamwork, communication, and problem-solving skills, which are essential for success in mathematics and beyond (Boaler, 2016).

3. Addressing Bias and Stereotypes:
   - By addressing bias and stereotypes in the mathematics classroom, teachers can create a more equitable and supportive learning environment that fosters the engagement and achievement of all students (Martin, 2009).

4. Preparing Students for the Future:
   - Collaborative and inclusive classroom practices prepare students for the collaborative and diverse workplaces of the future, where teamwork and cultural competence are valued skills (Department for Education, 2014).

Conclusion

Collaborative and inclusive classroom practices are essential for promoting equity, engagement, and achievement in mathematics education in the UK. By implementing practical strategies and addressing bias and stereotypes, teachers can create learning environments where all students feel valued, supported, and empowered to succeed.

References:

- Boaler, J. (2016). Mathematical Mindsets: Unleashing Students' Potential through Creative Mathematics, Inspiring Messages and Innovative Teaching. Jossey-Bass.
- Cobb, P., & Bowers, J. (1999). Cognitive and Situative Learning Perspectives in Theory and Practice. Educational Researcher, 28(2), 4-15.
- Department for Education. (2014). National Curriculum in England: Mathematics programmes of study.
- Gay, G. (2010). Culturally Responsive Teaching: Theory, Research, and Practice. Teachers College Press.
- Hattie, J. (2009). Visible Learning: A Synthesis of Over 800 Meta-Analyses Relating to Achievement. Routledge.
- Johnson, D. W., Johnson, R. T., & Smith, K. A. (2014). Cooperative Learning: Improving University Instruction by Basing Practice on Validated Theory. Journal on Excellence in College Teaching, 25(3&4), 85-118.
- Martin, D. B. (2009). Mathematics Success and Failure Among African-American Youth: The Roles of Sociohistorical Context, Community Forces, School Influence, and Individual Agency. Routledge.
- Rose, D. H., & Meyer, A. (2002). Teaching Every Student in the Digital Age: Universal Design for Learning. Association for Supervision and Curriculum Development (ASCD).
- Steele, C. M. (2010). Whistling Vivaldi: How Stereotypes Affect Us and What We Can Do. W.W. Norton

Implementing Change - Curriculum Design and Assessment

Understanding Change in Mathematics Education

Implementing change in mathematics education in the UK involves updating curriculum, teaching methods, and assessment practices to better meet the needs of diverse learners and address current challenges in the field. This includes considerations such as decolonising the curriculum, promoting inclusivity, and fostering continuous improvement through feedback and reflection.

Practicalities of Designing a Decolonised Maths Curriculum Framework

1. **Reviewing Current Curriculum**:
   - Conduct a thorough review of the existing maths curriculum to identify areas where colonial biases may be present. This involves examining curriculum content, resources, and assessment practices (Boaler, 2016).

2. **Incorporating Diverse Perspectives**:
   - Integrate contributions from diverse cultures and mathematicians into the curriculum. This includes highlighting non-Western mathematical traditions and their relevance to contemporary mathematics (Joseph, 2011).

3. **Developing Culturally Responsive Teaching Materials**:
   - Create teaching materials and resources that reflect the cultural backgrounds and experiences of students. Use examples, problems, and illustrations that resonate with diverse learners (Gay, 2010).

4. **Providing Professional Development**:
   - Offer training and support for teachers on decolonising the maths curriculum. This includes workshops, seminars, and resources that provide guidance on integrating diverse perspectives and promoting inclusivity (Nieto, 2010).

Inclusive Assessment Practices and Tools

1. **Mathematical Performance Tasks**:
   - Design performance tasks that allow students to demonstrate their mathematical understanding through real-world applications or creative projects. For example, students could design a budget for a family vacation, create a scale model of a city skyline using geometric shapes, or develop a board game that requires strategic thinking and mathematical reasoning.

2. **Interactive Digital Quizzes**:
   - Create interactive digital quizzes using platforms like Kahoot! or Quizzes that allow students to answer multiple-choice questions, solve problems, and engage in competitive gameplay. These quizzes can be customised to include multimedia

elements such as images, videos, and audio clips, making them accessible and engaging for diverse learners.

3. **Mathematical Inquiry Portfolios**:
   - Have students create mathematical inquiry portfolios where they document their investigations, explorations, and reflections on mathematical topics of interest. Portfolios can include written explanations, visual representations, and multimedia presentations that showcase students' mathematical thinking and problem-solving skills.

4. **Peer and Self-Assessment Rubrics**:
   - Develop peer and self-assessment rubrics that guide students in evaluating their own work and providing constructive feedback to their peers. Rubrics can include criteria such as mathematical accuracy, clarity of explanations, and use of mathematical vocabulary, empowering students to take ownership of their learning and development.

5. **Mathematical Modelling Projects**:
   - Assign mathematical modelling projects where students apply mathematical concepts to real-world problems and communicate their findings through written reports, presentations, or multimedia presentations. Projects can address issues relevant to students' lives and communities, fostering engagement and relevance in the assessment process.

6. **Collaborative Problem-Solving Tasks**:
   - Organise collaborative problem-solving tasks where students work in groups to tackle challenging mathematical problems or puzzles. Groups can use collaborative platforms like Google Jamboard or Padlet to brainstorm ideas, share strategies, and document their solutions collaboratively. This promotes teamwork, communication, and critical thinking skills.

7. **Mathematical Dialogue Journals**:
   - Implement mathematical dialogue journals where students engage in written conversations with their teacher or peers about their mathematical thinking, questions, and discoveries. Journals can be shared digitally or in print, allowing for asynchronous communication and reflection on mathematical concepts and problem-solving strategies.

8. **Mathematical Games and Simulations**:
   - Integrate mathematical games and simulations into the assessment process to provide students with interactive and engaging opportunities to apply their mathematical knowledge and skills. Games can be designed to address specific mathematical concepts or challenges, encouraging exploration and experimentation in a low-stakes environment.

9. **Digital Mathematics Portfolios**:
   - Create digital mathematics portfolios where students compile evidence of their mathematical learning and growth over time. Portfolios can include samples of students' work, reflections on their learning experiences, and self-assessments of

their mathematical proficiency. Digital portfolios allow for easy sharing and feedback from teachers and peers.

10. **Authentic Performance Assessments**:
    - Develop authentic performance assessments that require students to apply their mathematical understanding to real-world scenarios or tasks. For example, students could analyse data from a scientific experiment, design a statistical survey, or create a mathematical model to solve a practical problem. Authentic assessments promote relevance and transferability of mathematical skills to real-life situations.

11. **Multiple Assessment Methods**:
    - Use a variety of assessment methods to accommodate diverse learning styles and preferences. This includes formative assessments, project-based assessments, and performance tasks (Black & Wiliam, 1998).

12. **Authentic Assessment Tasks**:
    - Design assessment tasks that reflect real-world problem-solving scenarios and require students to apply mathematical concepts in context. This promotes deeper understanding and transfer of learning (Boaler, 2016).

13. **Flexible Assessment Criteria**:
    - Provide flexible assessment criteria that allow for different pathways to demonstrate proficiency. Consider students' individual strengths, interests, and cultural backgrounds when evaluating their performance (Tomlinson, 2014).

14. **Peer and Self-Assessment**:
    - Incorporate peer and self-assessment into the assessment process. Encourage students to evaluate their own work and provide constructive feedback to their peers. This promotes metacognitive skills and fosters a culture of collaboration (Hattie, 2009).

Mathematical Reflection Podcasts:

Have students create audio recordings or podcasts where they reflect on their mathematical learning experiences. Students can discuss mathematical concepts they found challenging, strategies they used to overcome obstacles, and connections they made between different topics. This multimedia approach allows students to express themselves verbally and share their reflections with their peers and teachers.

Virtual Reality Mathematics Tours:

Utilise virtual reality technology to create immersive mathematics experiences where students explore mathematical concepts in virtual environments. After the tour, students can reflect on their experience, discussing what they learned and how it impacted their understanding of the mathematical concepts. This interactive approach encourages engagement and deepens students' conceptual understanding.

Mathematical Inquiry Projects:

Encourage students to pursue inquiry-based projects where they investigate open-ended mathematical questions or problems of interest to them. Throughout the project, students can document their progress, record observations, and reflect on their learning journey. These projects provide opportunities for self-directed learning and promote critical thinking and problem-solving skills.

Mathematical Escape Rooms:

Design mathematical escape rooms or puzzle challenges where students must solve mathematics-related clues and puzzles to "escape" the room. After completing the activity, students can reflect on their problem-solving strategies, discussing what approaches were successful and what challenges they encountered. This hands-on, interactive approach makes learning mathematics engaging and memorable.

Mathematical Feedback Galleries:

Create a "mathematical feedback gallery" in the classroom where students display their work and receive feedback from their peers. Students can rotate through the gallery, providing written or verbal feedback on their classmates' work and reflecting on their own learning process. This peer-review process promotes collaboration, communication, and critical thinking skills.

Mathematical Thought Experiments:

Engage students in mathematical "thought experiments" where they explore hypothetical scenarios or conjectures related to mathematical concepts. After conducting the experiment, students can reflect on their findings, discussing the implications for their understanding of the mathematical principles involved. This approach encourages creative thinking and fosters curiosity and exploration.

Mathematical Feedback Puzzles:

Create feedback puzzles or challenges where students must solve mathematical problems to unlock clues or receive feedback on their work. Students can work individually or in groups to solve the puzzles, collaborating and communicating their reasoning along the way. This gamified approach to feedback makes the learning process fun and interactive.

Mathematical Performance Art:

Encourage students to express their mathematical understanding through performance art, such as skits, songs, or interpretive dance. After the performance, students can reflect on their creative process and how it deepened their understanding of mathematical concepts. This interdisciplinary approach integrates mathematics with the arts and promotes self-expression and creativity.

Continuous Improvement Through Feedback and Reflection

1. **Feedback Loops:**
   - Establish regular feedback loops between teachers and students to monitor progress and identify areas for improvement. Use both formal and informal feedback mechanisms to gather input from students about their learning experiences (Black & Wiliam, 1998).

2. **Reflective Practice:**
   - Encourage teachers to engage in reflective practice to evaluate the effectiveness of their teaching strategies and curriculum design. Provide opportunities for teachers to collaborate, share best practices, and learn from each other (Hattie, 2009).

3. **Data-Informed Decision Making:**

- Use data to inform instructional decisions and curriculum revisions. Analyse assessment results, student feedback, and classroom observations to identify trends and patterns that can inform future planning (Boaler, 2016).

4. **Continuous Professional Development**:
   - Offer ongoing professional development opportunities for teachers to enhance their knowledge and skills in mathematics education. This includes workshops, conferences, and online resources that focus on evidence-based practices and innovative approaches (Nieto, 2010).

Implementing change in mathematics education in the UK aligns with the goals and principles of the national maths curriculum by:

- **Promoting Equity and Inclusion**: Decolonising the curriculum and adopting inclusive assessment practices support the curriculum's aims of promoting equity, inclusion, and excellence in mathematics education for all students.
- **Fostering Continuous Improvement**: By fostering continuous improvement through feedback and reflection, teachers can ensure that the curriculum remains responsive to the evolving needs of students and the demands of society.

References:

- Black, P., & Wiliam, D. (1998). Inside the Black Box: Raising Standards Through Classroom Assessment. Phi Delta Kappan, 80(2), 139-148.
- Boaler, J. (2016). Mathematical Mindsets: Unleashing Students' Potential through Creative Mathematics, Inspiring Messages and Innovative Teaching. Jossey-Bass.
- Gay, G. (2010). Culturally Responsive Teaching: Theory, Research, and Practice. Teachers College Press.
- Hattie, J. (2009). Visible Learning: A Synthesis of Over 800 Meta-Analyses Relating to Achievement. Routledge.
- Joseph, G. G. (2011). The Crest of the Peacock: Non-European Roots of Mathematics. Princeton University Press.
- Nieto, S. (2010). The Light in Their Eyes: Creating Multicultural Learning Communities. Teachers College Press.
- Tomlinson, C. A. (2014). The Differentiated Classroom: Responding to the Needs of All Learners. ASCD.

Policy and Institutional Support - the Impact of Educational Policies

Educational policies play a crucial role in shaping the curriculum, pedagogy, and assessment practices in mathematics education. They provide guidelines and frameworks that influence what is taught, how it is taught, and how students' learning is assessed. In the context of decolonising mathematics education, educational policies can serve as catalysts for change by promoting inclusivity, diversity, and equity in the curriculum.

Practicalities of Designing a Decolonised Maths Curriculum Framework

1. **Policy Review and Revision**:
    - Conduct a comprehensive review of existing educational policies related to mathematics education to identify areas where colonial biases may be present. This involves examining curriculum standards, assessment frameworks, and teacher training requirements.

2. **Integration of Diverse Perspectives**:
    - Revise educational policies to explicitly include the integration of diverse cultural perspectives and contributions into the mathematics curriculum. This includes highlighting the achievements of mathematicians from non-Western cultures and incorporating examples and problems that reflect diverse cultural contexts.

3. **Professional Development**:
    - Allocate resources and support for teacher training and professional development programs that focus on decolonising mathematics education. Provide opportunities for teachers to learn about culturally responsive pedagogy, inclusive teaching practices, and strategies for integrating diverse perspectives into the curriculum.

4. **Collaborative Curriculum Development**:
    - Foster collaboration among educators, curriculum developers, researchers, and community stakeholders to design a decolonised maths curriculum framework. This collaborative approach ensures that diverse perspectives are represented and that the curriculum reflects the needs and interests of all students.

Examples of Successful Policy Implementation in the Mathematics Classroom

1. **New Zealand Mathematics Curriculum**:
    - The New Zealand mathematics curriculum incorporates Maori perspectives and cultural knowledge, recognising the contributions of indigenous communities to mathematics. It emphasises the importance of contextualising mathematics within

cultural and societal contexts, providing examples and problems that resonate with Maori students' lived experiences (Ministry of Education, New Zealand, 2007).

2. **Australian Curriculum: Mathematics**:
   - The Australian Curriculum: Mathematics includes cross-curricular priorities such as Aboriginal and Torres Strait Islander histories and cultures, sustainability, and Asia and Australia's engagement with Asia. It encourages teachers to explore the mathematical achievements and contributions of Indigenous Australians and other cultural groups, fostering a more inclusive and culturally responsive approach to mathematics education (Australian Curriculum, Assessment and Reporting Authority, 2016).

3. **United States Common Core State Standards for Mathematics**:
   - The Common Core State Standards for Mathematics in the United States emphasise the importance of mathematical practices such as reasoning, problem-solving, and communication. While not explicitly focused on decolonising mathematics education, these standards promote inclusive teaching practices that value diverse ways of thinking and problem-solving, providing a foundation for addressing cultural biases in the curriculum (Common Core State Standards Initiative, 2010).

Relevance to the UK Maths Curriculum

By revising policies to explicitly address the integration of diverse perspectives, providing professional development opportunities for teachers, and fostering collaborative curriculum development processes, policymakers can help create a more culturally responsive and inclusive mathematics curriculum that meets the needs of all students in England.

References:

- Australian Curriculum, Assessment and Reporting Authority. (2016). Australian Curriculum: Mathematics.

- Common Core State Standards Initiative. (2010). Common Core State Standards for Mathematics.

- Ministry of Education, New Zealand. (2007). The New Zealand Curriculum: Mathematics and Statistics.

Global Mathematics – Knowledge and ideas

One of the biggest issues that Mathematics teachers have, is the lack of knowledge about cultures around the world that can be used within their teaching to engage learners. Often universities and teacher training programmes omit non-European perspectives within their programmes, thus leaving teachers barren of knowledge and understanding of global mathematics and how to teach and influence diverse classrooms to create mathematicians for the future. Where learners are taught to see mathematics in the world around them and be curious enough to seek out, identify solve and develop concepts based on this new world.

When was the last time a learner described your mathematics lesson/lecture/workshop as eye-opening, awesome, or intriguing? What was it like to see/hear/feel their excitement?

The following section will provide you with knowledge and ideas for you to investigate further for your lessons:

Geometric Patterns in Architecture:

West African architecture often incorporates intricate geometric patterns. One such pattern is the "labyrinth" design, commonly found in mosques and buildings. This pattern consists of interlocking squares, creating a visually appealing mase-like structure. Teachers can introduce this pattern to students and explore its properties, such as the number of squares and their arrangement. Students can analyse and reproduce these patterns using geometric principles, including transformations, symmetry, and tessellations.

1. Geometric Patterns: West African cultures, such as the Ashanti and Yoruba, have a long history of creating intricate geometric patterns using simple tools and materials. These patterns often feature symmetrical designs and repetitive motifs.

Equation: One example is the Sankofa symbol from the Akan people, which can be represented by the equation:

$$A = r \sin(\theta)$$

Explanation: The Sankofa symbol consists of a bird with its head turned backward, symbolising the importance of learning from the past. The equation above represents a polar equation, where "r" represents the distance from the origin, and "θ" represents the angle. By exploring this equation and creating geometric patterns based on it, students can understand the relationship between polar coordinates and geometric shapes.

Application: Students can study the Sankofa symbol and create their own geometric patterns inspired by West African designs. This activity can support the UK national curriculum's objectives related to geometry, symmetry, and transformations.

2. Number Systems: West Africa has a history of using various number systems, such as the Yoruba's "Aroko" and the Igbo's "Nsimbu." These number systems often involve counting in base-20 or base-12, reflecting cultural practices and practical needs.

Equation: Let's consider the Yoruba Aroko number system. In Aroko, the equation for counting in base-20 can be represented as follows:

$n = 20a + b$

Explanation: In the Aroko system, "a" represents the number of twenties, and "b" represents the remaining units. This system provides a unique perspective on place value and can broaden students' understanding of different number bases.

Application: Students can explore the Aroko number system and compare it to the decimal system used in the UK. They can convert numbers between the two systems, investigate patterns, and recognise the cultural influence on number systems. This activity can help students develop their number sense and deepen their understanding of place value.

3. Ethnomathematics: Ethnomathematics is the study of mathematical knowledge and practices within different cultures. It highlights the cultural context and relevance of mathematics.

Equation: Rather than a specific equation, ethnomathematics focuses on the relationships between mathematics and cultural practices. For example, students can examine the calculations involved in traditional African weaving patterns, such as the Kente cloth from Ghana, and explore the geometric principles underlying these designs.

Explanation: By studying the mathematical principles embedded in cultural artifacts like Kente cloth, students can understand how mathematics is interwoven into everyday practices. They can explore concepts such as symmetry, tessellations, and rotational symmetry that are essential to the creation of these intricate designs.

Application: Students can engage in hands-on activities related to ethnomathematics, such as creating their own patterns inspired by Kente cloth or other African designs. This approach allows students to explore mathematical concepts in a cultural context and promotes a deeper appreciation for the diverse applications of mathematics.

Fractals in Adinkra Symbols:

Adinkra symbols, originating from Ghana, are visual representations of concepts and proverbs. Many of these symbols exhibit fractal-like properties, with intricate self-replicating patterns. Teachers can introduce Adinkra symbols and discuss their recursive nature. Students can explore self-similarity by creating their own Adinkra-inspired fractal designs using repeated transformations, such as scaling and translation. They can also investigate the relationship between the scale factor and the number of iterations to deepen their understanding of fractals.

1. Fractal Geometry: Fractals are intricate geometric patterns that repeat at different scales. The concept of fractals has been used in West African traditional art, such as the patterns found in adinkra symbols.

Equation: The Sierpinski Triangle is a famous fractal pattern that can be generated using a recursive equation:

$T(n) = T(n-1) + T(n-1)$

Explanation: In the Sierpinski Triangle, T(n) represents the number of triangles at the nth iteration. The equation states that to generate the next iteration, you take the previous iteration and add another copy of itself.

Application: Students can explore fractal geometry by studying the Sierpinski Triangle and other fractal patterns found in West African art. They can learn about recursion, self-similarity, and the concept of infinity. This activity aligns with the GCSE Mathematics curriculum's objectives related to geometry, sequences, and patterns.

2. Number Bases: West Africa has a history of using different number systems, such as the base-20 system used by the Yoruba people. Understanding alternative number bases can deepen students' understanding of place value and broaden their perspective on mathematical systems.

Equation: Let's consider the base-20 system used by the Yoruba people. In this system, counting can be represented by the following equation:

$n = 20a + b$

Explanation: In the Yoruba base-20 system, "a" represents the number of twenties, and "b" represents the remaining units. This equation demonstrates the concept of place value and provides an opportunity for students to explore different number bases.

Application: Students can investigate the Yoruba base-20 system and compare it to the decimal system used in the UK. They can convert numbers between the two systems, explore patterns, and recognise the cultural influence on number systems. This activity supports the GCSE Mathematics curriculum's objectives related to number systems, place value, and number operations.

3. Traditional Measurement Systems: West African cultures have developed unique measurement systems based on everyday objects and body parts. These systems provide practical and culturally relevant ways of measuring quantities.

Equation: An example is the traditional African measurement system using the length of a finger segment as a unit of measurement. An equation could be:

$L = nf$

Explanation: In this equation, "L" represents the length being measured, "n" represents the number of finger segments, and "f" represents the length of a finger segment.

Application: Students can explore traditional African measurement systems and their relationship to the decimal metric system. They can compare measurements using finger segments, hand spans, or other traditional units with the metric measurements they are familiar with. This activity supports the GCSE Mathematics curriculum's objectives related to units of measurement, conversions, and proportional reasoning.

Sona Symbolic Writing System:

The Sona writing system, developed by the Bamum people of Cameroon, represents mathematical ideas through symbols. Each symbol corresponds to a mathematical concept, such as addition, subtraction, multiplication, or division. Teachers can introduce the Sona writing system and discuss its symbolic representations. Students can practice translating mathematical expressions into Sona symbols and vice versa, reinforcing their understanding of mathematical operations.

Here is one way you can introduce this to a class:

Today, we will be exploring a fascinating writing system called Sona, used by the Bambara people. Sona symbols are not just letters or words but rather symbolic representations that convey meaning. What makes Sona unique is that it can be used to represent not only language but also mathematical expressions. This connection between language and mathematics allows us to delve deeper into the world of symbols and explore the beauty of mathematics through a cultural lens.

Let's begin by discussing the Sona writing system and its symbolic representations. In the Sona system, each symbol represents a specific concept or idea. These symbols are composed of simple geometric shapes, such as lines, circles, squares, and triangles. By combining these shapes in different ways, the Bambara people were able to create a rich vocabulary of symbols.

One remarkable aspect of Sona symbols is their ability to represent mathematical expressions. Just as we use numbers and symbols in mathematical equations, Sona symbols can be used to represent mathematical operations. For example, a circle may represent addition, a square might symbolise multiplication, and a triangle could stand for division. By associating these geometric shapes with mathematical operations, the Bambara people created a unique way of expressing mathematical concepts.

Sona Symbols Exercise: Now, let's put our understanding of Sona symbols and mathematical expressions to the test. I will give you some mathematical expressions, and your task is to translate them into Sona symbols. Then, we will reverse the process and translate Sona symbols back into mathematical expressions.

1. Mathematical Expression: 2 + 3 * 4 Translate this expression into Sona symbols.
2. Mathematical Expression: (6 - 2) / 3 Translate this expression into Sona symbols.
3. Sona Symbol: ○ △ ☐ Translate this Sona symbol into a mathematical expression.
4. Sona Symbol: ☐ ○ △ Translate this Sona symbol into a mathematical expression.

Take a few minutes to work on these exercises independently. Once you have completed them, we will discuss our translations and explore the connections between Sona symbols and mathematical expressions.

(Allow time for students to complete the exercises.)

Discussion: Now, let's go over our translations and discuss the connections between the Sona symbols and mathematical expressions.

For the first exercise, the mathematical expression "2 + 3 * 4" can be translated into Sona symbols as ○ + △ ☐ .

In the second exercise, the mathematical expression '(6 - 2) / 3" can be translated into Sona symbols as △ - ○ ☐ .

For the third exercise, the Sona symbol ○ △ ☐ can be translated back into a mathematical expression as "2 * 3 + 4".

Lastly, the Sona symbol ▢ ○ △ can be translated into a mathematical expression as "4 + 2 * 3".

Through these exercises, we can see how Sona symbols can represent mathematical operations and expressions. By exploring the connection between language and mathematics, we gain a deeper appreciation for the power of symbols in expressing complex ideas.

Conclusion: Today, we have delved into the intriguing world of the Sona writing system. We have discussed its symbolic representations, explored its connection to mathematical expressions, and even practiced translating between Sona symbols and mathematical notation. By studying the Sona system, we have expanded our understanding of symbols, language, and mathematics, while also appreciating the cultural heritage of the Bambara people.

What else could we do with the Sona symbolic writing system?

Discuss how these symbols are composed of geometric shapes and patterns. Examine various Sona symbols and help students identify the geometric shapes present in each symbol. Encourage them to recognise common shapes like circles, triangles, squares, rectangles, and hexagons.

Discuss the concept of symmetry in Sona symbols. Identify lines of symmetry within the symbols and explore how they divide the design into equal halves. Emphasise both vertical and horizontal lines of symmetry.

Engage students in creating their own Sona symbols using geometric shapes. Encourage them to incorporate symmetry in their designs and discuss the mathematical reasoning behind their choices.

Conduct a Sona symbol symmetry hunt. Provide students with a collection of Sona symbols, and ask them to identify and classify the symbols based on their symmetry properties. Discuss the different types of symmetry found within the symbols.

Introduce the concept of transformations, such as reflection, rotation, and translation, using Sona symbols. Have students experiment with transforming Sona symbols by reflecting or rotating them and discussing how these transformations affect symmetry.

Engage students in creating artwork inspired by Sona symbols and symmetry. Provide them with a variety of geometric shapes and challenge them to design symmetrical compositions using the principles observed in Sona symbols.

Explore the stories and meanings associated with Sona symbols. Discuss how the use of specific shapes and symmetry in the symbols conveys cultural and symbolic significance. Encourage students to analyse the relationship between the symbolism and the geometric representation.

Create puzzles or challenges using Sona symbols to reinforce geometric concepts and symmetry. These can include activities like completing partial Sona symbols by reflecting or rotating existing portions to create symmetry.

Encourage students to maintain reflection journals where they record their thoughts, observations, and discoveries about Sona symbols and their connection to geometry and symmetry. This provides an opportunity for self-expression and deepening their understanding.

By exploring Sona symbols in the English classroom, students can develop a deeper appreciation for geometry and symmetry while also gaining cultural awareness and understanding. These activities promote critical thinking, creativity, and mathematical reasoning within a culturally rich context.

Akan Fibonacci-Like Number Sequences

Akan artisans have long incorporated complex mathematical patterns into their designs, including sequences that reflect growth and proportion similar to what is now known as the Fibonacci sequence. These traditions in weaving and design date back to the 12th century and precede the formal documentation of the Fibonacci sequence by European mathematicians. The now named Fibonacci sequence is a series of numbers in which each number is the sum of the two preceding ones (e.g., 0, 1, 1, 2, 3, 5, 8, 13, 21, and so on). The Akan people, renowned for their exquisite craftsmanship, have embedded these sequences into their cultural practices for centuries.

Akan Weaving and Kente Cloth

One prominent example of these mathematical principles can be seen in the weaving of traditional Akan Kente cloth. This fabric is celebrated for its vibrant colours and intricate geometric motifs. The patterns often include sequences of rectangles that mirror the proportions found in the Fibonacci sequence. The widths of these rectangles align with numbers in the sequence, resulting in designs that are visually harmonious and balanced. This use of proportionality not only enhances the aesthetic appeal of the cloth but also reflects an innate sense of natural order.

Jewellery Design

Akan artisans also apply similar sequences in their jewellery designs. Beads and stones are arranged with precision, often in patterns that mirror these growth sequences. For instance, a necklace may feature sections with a number of beads corresponding to the Fibonacci sequence. This meticulous arrangement enhances the jewellery's visual appeal, creating a sense of harmony and balance that is both visually and culturally significant.

Architecture

The principles of these sequences extend to Akan architecture as well. The layout and proportions of buildings, from the placement of windows and doors to the dimensions of interior spaces, often follow ratios that are akin to those in the Fibonacci sequence. This adherence to proportionality creates structures that are visually harmonious and unified, reflecting a deep understanding of balance and growth.

Mathematical Sophistication

The use of these sequences in Akan art and design highlights the artisans' profound understanding of mathematical principles and their ability to create beauty through order and proportion. By incorporating these ancient patterns, Akan artisans have crafted designs that resonate with natural harmony. This tradition continues to influence contemporary Akan art, preserving the rich cultural heritage and mathematical sophistication of the Akan people for future generations to admire.

In essence, Akan artisans have long utilised number sequences akin to the Fibonacci sequence to create designs that are both aesthetically pleasing and mathematically grounded. This tradition demonstrates the enduring creativity and intellectual depth of Akan craftsmanship, showcasing their contributions to art and mathematics well before such concepts were formalised in the Western world.

But we don't stop there, we can also use the everyday hobbies of our learners to enthuse…

Global Maths – Mathematics without limits

Algebra: Linear Equations

Concept: Solving Linear Equations

Example: Video Game Economics

Classroom Breakdown:

- Game: Use popular video games like *FIFA* or *Fortnite*, which have in-game economies where players buy and sell items or players using virtual currency.

- Scenario: Suppose a player in *FIFA* wants to buy a player card costing 2000 coins. They earn 150 coins per match they play.

- Math Problem: How many matches do they need to play to afford the player card?

- Equation: Let x be the number of matches. The equation is $150x = 2000$.
- Solution:

$$150x = 2000$$

$$x = \frac{2000}{150} \approx 13.33$$

The player needs to play 14 matches (rounding up since they can't play a fraction of a match).

- Discussion: Discuss how real-world economics in games mirrors budgeting and saving in real life.

Geometry: Angles and Trigonometry

Concept: Angles of Elevation and Depression

Example: Film and Photography

Classroom Breakdown:

- Film: Discuss the use of camera angles in a British film like *Skyfall* or *Dunkirk*.

- Scenario: Calculate the angle of elevation for a camera to capture an overhead shot of a subject 3 meters tall, where the camera is 5 meters away horizontally.

- Diagram: Draw a right triangle where the opposite side is 3 meters, and the adjacent side is 5 meters.

- Maths Problem: Calculate **the angle of elevation.**
- Trigonometry:

$$\tan(\theta) = \frac{\text{opposite}}{\text{adjacent}} = \frac{3}{5}$$

$$\theta = \tan^{-1}\left(\frac{3}{5}\right) \approx 30.96°$$

- Activity: Students can recreate this in class with a camera or smartphone and measure angles using apps.
- Discussion: Talk about how different angles affect perception in movies and photography.

Statistics: Data Analysis

Concept: Measures of Central Tendency

Example: Music Charts Analysis

Classroom Breakdown:

- Music: Use data from the UK Top 40 music charts.
- Scenario: Analyse the lengths of songs in the top 10 to calculate the mean, median, and mode of song lengths.
- Data Collection: List song durations (in minutes) for analysis.
- Example Data: 3.5, 4.0, 3.2, 4.5, 3.5, 4.1, 3.6, 4.2, 3.9, 3.8
- Mean Calculation:

$$\text{Mean} = \frac{3.5 + 4.0 + 3.2 + 4.5 + 3.5 + 4.1 + 3.6 + 4.2 + 3.9 + 3.8}{10} = 3.83 \text{ minutes}$$

- Median Calculation: Order the data: 3.2, 3.5, 3.5, 3.6, 3.8, 3.9, 4.0, 4.1, 4.2, 4.5

    Median = $\frac{3.8+3.9}{2} = 3.85$ minutes

- Mode: 3.5 (occurs twice)
- Discussion: Discuss how song length affects popularity and listener engagement.

Probability: Experimental Probability

Concept: Probability of Events

Example: Sports Betting and Outcomes

Classroom Breakdown:

- Sport: Use data from football matches in the Premier League.
- Scenario: Calculate the probability of a team winning given their past performance.
- Data Collection: Use a team's past 10 matches: Wins = 6, Draws = 2, Losses = 2.
- Probability Calculation:
- Probability of winning = $\frac{\text{Number of Wins}}{\text{Total Matches}} = \frac{6}{10} = 0.6$
- Probability of drawing = $\frac{2}{10} = 0.2$
- Probability of losing = $\frac{2}{10} = 0.2$
- Simulation Activity: Create a class simulation of a football league, with students calculating probabilities and making predictions.
- Discussion: Discuss how statistics inform betting and team strategies.

Calculus (A-Level): Rates of Change

Concept: Derivatives and Motion

Example: Video Game Motion Analysis

Classroom Breakdown:

- Game: Analyse character movement in a racing game like *Forza Horizon*.
- Scenario: Calculate the acceleration of a car given its velocity-time graph.
- Graph Interpretation: Provide a velocity-time graph showing linear acceleration.
- Derivative Calculation:
  - If velocity $v(t) = 4t + 2$, find the acceleration.
  - $a(t) = \frac{dv}{dt} = \frac{d}{dt}(4t + 2) = 4\,\text{m/s}^2$
- Activity: Use the graph to predict and analyse different motion scenarios in the game.
- Discussion: Discuss how calculus is used in game development to simulate realistic physics.

6. Ratio and Proportion

Concept: Proportional Relationships

Example: Cooking with Ratios in Music Videos

Classroom Breakdown:

- Music Video: Analyse a popular British cooking-themed music video, like Ed Sheeran's "Take Me Back to London."

Global Maths – Mathematics without limits

- Scenario: Calculate ingredient ratios to scale a recipe featured in the video.
- Example Recipe: Original for 4 people: 200g flour, 100g sugar, 2 eggs.
- Scaling Problem: Scale recipe for 10 people.
- Ratio Calculation:

$$\text{Flour: } \frac{200}{4} \times 10 = 500\,\text{g}$$

$$\text{Sugar: } \frac{100}{4} \times 10 = 250\,\text{g}$$

$$\text{Eggs: } \frac{2}{4} \times 10 = 5\,\text{eggs}$$

Activity: Students scale recipes and discuss cultural dishes from their backgrounds

- Discussion: Explore how ratios are vital in cooking, music production, and video editing.

Graph Theory

Concept: Network Analysis

Example: Social Media Networks

Classroom Breakdown:

- Network: Use the structure of a social media platform like Instagram.
- Scenario: Model friendships and connections as a graph, where nodes represent users and edges represent connections.
- Maths Problem: Calculate the shortest path between two users using Dijkstra's algorithm.
- Graph Representation: Create a sample network with weighted edges representing the strength of relationships.
- Algorithm Application:
    - Assign weights to edges based on frequency of interactions.
    - Use Dijkstra's algorithm to find the shortest path.
- Activity: Have students map their own social networks and analyse connectivity.
- Discussion: Discuss how companies use graph theory for targeted advertising and user experience optimisation.

Concept: Quadratic Equations

Mathematical Breakdown:

- **Quadratic Formula:** Solve quadratic equations using the formula: $ax^2 + bx + c = 0$ with $x = \frac{-b \pm \sqrt{b^2 - 4ac}}{2a}$.
- **Factoring:** Factor quadratic expressions into the form $(x - p)(x - q) = 0$.
- **Completing the Square:** Transform the equation into a perfect square form: $(x - h)^2 = k$.

Cultural Approach:

- **Cultural Patterns:**
  - **Example:** Explore traditional geometric patterns in Islamic art or African textiles by creating quadratic equations that describe symmetrical patterns.
  - **Activity:** Have students bring in a pattern from their cultural background. Use these patterns to model parabolas or symmetrical designs, discussing the role of symmetry and curves described by quadratic equations.

- **Agricultural Contexts:**
  - **Example:** Discuss optimisation problems, such as maximising crop yield in a given area, using quadratics to determine the best dimensions for planting.
  - **Activity:** Create a project where students calculate the optimal dimensions of a rectangular plot to maximise area or perimeter using quadratic equations.

- **Personal Finances:**
    - **Example:** Use quadratics to calculate profits and losses in a simple business model, such as a lemonade stand or small shop.
    - **Activity:** Have students model revenue ($R(x) = px$) and cost ($C(x) = ax^2 + bx + c$) equations, and find the break-even points and profit maximization by solving $R(x) = C(x)$.

Geometry and Trigonometry

Concept: Properties of Shapes and Trigonometry

Mathematical Breakdown:

- **Pythagorean Theorem:** $a^2 + b^2 = c^2$ for right triangles.
- **Trigonometric Ratios:** $\sin(\theta) = \frac{opposite}{hypotenuse}$, $\cos(\theta) = \frac{adjacent}{hypotenuse}$, $\tan(\theta) = \frac{opposite}{adjacent}$.

Cultural Approach:

- **Architectural Heritage:**

- o **Example:** Analyse the geometric design of the Great Mosque of Córdoba or the symmetry in the design of the Taj Mahal using theorems and trigonometry.
- o **Activity:** Assign projects where students calculate angles, areas, and dimensions of cultural monuments using trigonometric identities and geometric formulas.
- **Navigation and Astronomy:**
  - o **Example:** Discuss how ancient sailors used trigonometry for navigation based on the stars.
  - o **Activity:** Use historical navigation problems to calculate distances and angles, applying trigonometric concepts.

Statistics and Probability

Concept: Data Analysis and Probability

Mathematical Breakdown:

- **Mean, Median, Mode:** Calculate measures of central tendency for data sets.
- **Probability Calculations:** Use probabilities in real-world contexts, like sports outcomes or social issues.

Cultural Approach:

- **Sports Statistics:**
  - o **Example:** Use data from football or cricket matches to calculate averages and probabilities of outcomes.
  - o **Activity:** Assign projects where students collect and analyse statistics from their favourite sports, creating probability models for future matches.
- **Social Justice and Demographics:**
  - o **Example:** Analyse demographic data related to racial equity, health outcomes, or income distribution.
  - o **Activity:** Have students use real data to discuss and present findings on social issues, using graphs and statistical measures to support their arguments.

Calculus (A-Level)

Concept: Limits, Differentiation, Integration

Mathematical Breakdown:

- **Limits:** Understand behaviour of functions as inputs approach a value.

- **Differentiation:** Find derivatives to understand rates of change, slopes of tangents.
- **Integration:** Calculate area under curves, total accumulation of quantities.

Cultural Approach:

- **Engineering and Technology:**
    - **Example:** Discuss how calculus is used in designing safety features in cars, an industry many students might find aspirational.
    - **Activity:** Engage students in a project where they use differentiation to find rates of change in vehicle speed, or use integration to calculate fuel consumption over time.
- **Biology and Medicine:**
    - **Example:** Discuss models of population growth or spread of diseases using calculus.
    - **Activity:** Use differential equations to model real-world scenarios such as disease spread or population dynamics in a biology project.

Number Theory

Concept: Prime Numbers, Divisibility, Patterns

Mathematical Breakdown:

- **Prime Numbers:** Understand properties, find prime numbers, use in cryptography.
- **Divisibility Rules:** Apply rules to solve problems efficiently.

Cultural Approach:

- **Cryptography:**
    - **Example:** Explain how prime numbers secure digital communications.
    - **Activity:** Have students use simple encryption algorithms, like RSA, to encode and decode messages, highlighting the practical use of primes.
- **Cultural Mathematics:**
    - **Example:** Explore historical number systems, such as the Mayan base-20 or African tally systems.
    - **Activity:** Assign a project where students research and present on the number systems from their heritage and their applications in ancient commerce or architecture.

Vectors and Matrices (A-Level)

Concept: Vector Operations and Matrix Transformations

Mathematical Breakdown:

- **Vector Operations:** Perform addition, subtraction, scalar multiplication, dot product.
- **Matrices:** Use for transformations, solve systems of equations.

Cultural Approach:

- **Physics and Engineering:**
    - **Example:** Show how vectors are used to model forces in engineering projects.
    - **Activity:** Have students calculate forces acting on structures like bridges, connecting to local engineering projects.
- **Art and Graphics:**
    - **Example:** Use vectors to create digital art, showing transformations through matrix operations.
    - **Activity:** Engage students in a project to design a piece of digital art using vector graphics software, highlighting the use of transformations.

Ratio and Proportion

Concept: Solving Problems with Ratios, Scaling, Rates

Mathematical Breakdown:

- **Ratios and Proportions:** Use to solve scaling problems, understand direct and inverse relationships.

Cultural Approach:

- **Culinary Arts:**
    - **Example:** Use recipes to teach ratios and scaling, encouraging students to explore dishes from their cultures.
    - **Activity:** Organise a "mathematics in the kitchen" day where students modify recipes to serve different numbers of people, calculating new ingredient amounts using ratios.
- **Music and Rhythm:**
    - **Example:** Explore rhythmic patterns in African drumming or South Asian classical music to illustrate mathematical ratios.
    - **Activity:** Have students create and perform musical pieces based on mathematical rhythm patterns, linking to music theory.

Video Game Design and Programming

Math Concept: Coordinate Geometry

Teaching Level: Secondary School (KS3/KS4)

Classroom Breakdown:

- Hobby/Activity: Many students enjoy video games, but they can also be interested in how games are designed and developed.

- Link to Math: Use coordinate geometry to explain how characters move in a game environment.

- Teaching Approach:

    o Introduction to Coordinate Geometry: Explain the concept of the coordinate plane (x and y axes). Use graph paper or an interactive whiteboard to show how positions are plotted.

    o Activity: Create a simple grid-based game where students plot points to move a character (e.g., move from (2, 3) to (5, 8)).

    o Real-World Application: Discuss how game developers use coordinates to place objects and characters in a virtual world.

    o Extension: Use basic coding platforms like Scratch or Python to allow students to create simple games where they implement coordinate-based movement.

- Outcome: Students understand the practical application of coordinates in technology and can visualise how math is used in game design.

Music Production

Math Concept: Ratios and Proportions

Teaching Level: Primary and Secondary School (KS2/KS3)

Classroom Breakdown:

- Hobby/Activity: Many students are interested in music production, whether playing instruments or creating digital music tracks.

- Link to Math: Explore the concept of rhythm, beat, and timing in music using ratios and proportions.

- Teaching Approach:

    o Introduction to Ratios: Explain the concept of ratios using simple examples, such as comparing the number of beats in different sections of a song (e.g., a 4/4 time signature).

    o Activity: Have students create a simple rhythm using classroom instruments or music software. Ask them to create patterns that follow specific ratios (e.g., 2:1 or 3:2).

- Discussion: Discuss how different music styles use various time signatures and rhythms, highlighting how ratios and proportions play a crucial role.
- Extension: Use digital audio workstations (DAWs) like GarageBand or Ableton to create music loops and explore how altering ratios affects the music.
- Outcome: Students understand how mathematical ratios are essential in creating and understanding music, reinforcing the connection between math and artistic expression.

Robotics and Artificial Intelligence

Math Concept: Algorithms and Sequences

Teaching Level: Secondary School (KS3/KS4)

Classroom Breakdown:

- Hobby/Activity: Robotics clubs and competitions are becoming increasingly popular, with students designing and programming robots.
- Link to Math: Teach algorithms and sequences through robot programming and task execution.
- Teaching Approach:
    - Introduction to Algorithms: Explain algorithms as step-by-step instructions to complete a task. Use simple examples like following a recipe.
    - Activity: Use educational robots like LEGO Mindstorms or micro

. Have students program the robot to complete a series of tasks using a sequence of instructions.

    - Real-World Application: Discuss how robots use algorithms in industries, such as manufacturing or autonomous vehicles.
    - Extension: Challenge students to optimise their algorithms for efficiency, introducing concepts of optimisation and iteration.
- Outcome: Students learn how algorithms are the foundation of programming and robotics, providing insight into the technologies of the future.

Social Media Analytics

Math Concept: Data Analysis and Statistics

Teaching Level: Secondary School (KS3/KS4)

Classroom Breakdown:

- Hobby/Activity: Social media is a significant part of young people's lives. Understanding how data is used in this context can be engaging.
- Link to Math: Use data analysis and statistics to interpret social media trends.
- Teaching Approach:
    - Introduction to Data Analysis: Explain data collection, representation, and interpretation using simple datasets.
    - Activity: Provide students with a dataset representing social media usage statistics (e.g., average screen time, engagement rates).
    - Analysis: Have students calculate measures of central tendency (mean, median, mode) and create graphs to represent the data visually.
    - Discussion: Discuss how companies use data to make decisions about advertising and content creation, highlighting ethical considerations.
    - Extension: Introduce basic concepts of predictive analytics and how trends can be forecasted.
- Outcome: Students develop skills in data analysis and understand its application in social media, a field relevant to their daily lives.

Environmental Conservation

Math Concept: Measurement and Geometry

Teaching Level: Primary and Secondary School (KS2/KS3)

Classroom Breakdown:

- Hobby/Activity: Environmental awareness and conservation efforts are significant interests among young people today.
- Link to Math: Use measurement and geometry to understand and solve real-world environmental problems.
- Teaching Approach:
    - Introduction to Measurement and Geometry: Explain basic geometric shapes and how they can be used to measure and calculate areas and volumes.
    - Activity: Have students measure the school garden or a local park to calculate areas for planting or conserving green spaces.
    - Real-World Application: Discuss how geometry and measurement are used in environmental planning, such as designing efficient land use.
    - Extension: Introduce concepts like calculating carbon footprints or optimising resource usage using geometry.
- Outcome: Students see how math is applied in environmental conservation, fostering an understanding of how math can be used to impact the world positively.

Fashion Design

Math Concept: Geometry and Proportionality

Teaching Level: Secondary School (KS3/KS4)

Classroom Breakdown:

- Hobby/Activity: Fashion design is an area of interest that can be linked to various math concepts, particularly geometry.
- Link to Math: Use geometry and proportions to design clothing patterns.
- Teaching Approach:
    - Introduction to Geometry in Design: Explain how geometric shapes are used to create patterns in fashion.
    - Activity: Have students design a simple clothing pattern using basic geometric shapes. Use paper or fabric to cut and assemble their designs.
    - Discussion: Explore how designers use proportions to ensure a good fit and how math is essential in the production process.
    - Extension: Discuss the use of symmetry and tessellation in pattern design, integrating these concepts into students' projects.
- Outcome: Students learn how geometry is foundational in fashion design, linking math to creative and practical applications.

Virtual Reality (VR) Experiences

Math Concept: 3D Geometry and Vectors

Teaching Level: Secondary School (KS4/A-Level)

Classroom Breakdown:

- Hobby/Activity: Virtual reality is an emerging field that fascinates many students, offering immersive experiences.
- Link to Math: Use 3D geometry and vectors to understand VR environments.
- Teaching Approach:
    - Introduction to 3D Geometry and Vectors: Explain basic 3D shapes, vector addition, and how these concepts are used in creating VR environments.

- o Activity: Use VR software or tools like Blender to allow students to create simple 3D models. Discuss how vectors are used to position and move objects within these environments.
- o Discussion: Explore how VR developers use math to create realistic experiences and how vectors are integral to programming motion and interaction.
- o Extension: Introduce concepts like transformations and matrix operations used in advanced VR development.
- Outcome: Students gain an understanding of how advanced math is applied in cutting-edge technologies, encouraging interest in STEM fields.

Quadratic Equations in Personal Finance

Mathematical Concept: Quadratic Equations

Teaching Level: Secondary School (KS4)

How to Teach:

- **Concept Overview:** Introduce quadratic equations in the standard form $ax^2 + bx + c = 0$.
- Explain how these equations can model real-world scenarios, particularly in finance.
- Cultural Context: Many students and families manage budgets, save money, and make purchasing decisions. Use this context to create meaningful problems.
- Example Problem: Suppose a student wants to buy a new gaming console, which they plan to finance by selling a collection of old games and doing part-time work. Model the profit scenario using quadratic equations.
- Teaching Steps:
    1. Introduce the Scenario:
        - Scenario: A student sells games and earns £10 for each game. They also do part-time work, earning £50 per week. They need £300 to buy a console.
        - Objective: Find out how many weeks they need to work to reach their goal if they sell xxx games.

- Formulate the Quadratic Equation:
    - Total money earned = $10x + 50w = 300$.
    - Suppose each week, they can sell 2 games, leading to a model:
    $$10(2w) + 50w = 300$$
    - Simplify and form a quadratic equation:
    $$20w + 50w = 300 \Rightarrow 70w = 300 \Rightarrow w^2 + w - 15 = 0$$
  3. **Solve the Quadratic Equation:**
     - Use the quadratic formula $w = \frac{-b \pm \sqrt{b^2 - 4ac}}{2a}$.
     - Here, $a = 1, b = 1, c = -15$.
     - Calculate:
     $$w = \frac{-1 \pm \sqrt{1 + 60}}{2} = \frac{-1 \pm \sqrt{61}}{2}$$
     - Estimate or calculate exact weeks needed.
  - Discuss Results:
    - Discuss how this calculation affects decision-making in personal finance.
    - Reflect on the real-world implications and the importance of planning.
- Outcome: Students understand quadratic equations' applications in finance and gain problem-solving skills relevant to everyday decisions.

Trigonometry in Music and Sound Waves

Mathematical Concept: Trigonometric Functions

Teaching Level: Secondary School (KS4/A-Level)

How to Teach:

- Concept Overview: Introduce sine and cosine functions, and how they model periodic phenomena like sound waves.
- Cultural Context: Music is universal, and many students engage with it regularly. Different cultures have distinct musical styles that can be analysed mathematically.
- Example Problem: Analyse the sound wave patterns of traditional African drumming or Caribbean reggae rhythms using trigonometric functions.
- Teaching Steps:
  1. Introduce the Concept:
     - Explain how sound waves are modelled by sine and cosine functions.
     - Define key parameters: amplitude, frequency, and phase shift.
  2. Explore Cultural Music Patterns:

- Use software or online tools to visualise sound waves of traditional music from students' backgrounds.
  - Example: Plot the sound wave of a drumbeat as $y = A\sin(Bx + C)$.

3. Analyse the Waveform:
   - Identify amplitude (A), which reflects volume, and frequency (B), which indicates pitch.
     - Example: For a drumbeat with a peak amplitude of 5 units and a frequency corresponding to 3 beats per second:
     $$y = 5\sin(3x)$$
4. Calculate and Interpret:
   - Have students calculate different parameters for given musical pieces.
   - Discuss how changes in frequency and amplitude affect the music and cultural expression.
5. Create a Cultural Connection:
   - Relate the mathematical patterns to cultural music traditions and their significance.

- Outcome: Students appreciate the mathematical basis of music, understand trigonometric functions, and explore cultural connections.

Statistics in Sports Performance

Mathematical Concept: Probability and Statistics

Teaching Level: Secondary School (KS3/KS4)

How to Teach:

- Concept Overview: Introduce mean, median, mode, and probability concepts to Analyse sports statistics.
- Cultural Context: Sports are a major interest for many students. They can relate statistical analysis to their favourite teams or athletes.
- Example Problem: Analyse a dataset of a cricket player's performance (popular in South Asian communities) or a footballer's scoring record.
- Teaching Steps:
  1. Collect and Present Data:
     - Use real-world data from sports databases.

- Example: Analyse the number of runs scored by a cricket player in a season.
    2. Calculate Central Tendencies:
        - Mean: Average runs scored per match.
        $$\text{Mean} = \frac{\text{Total runs}}{\text{Number of matches}}$$
        - Median: The middle value of runs scored.
        - Mode: The most frequently occurring score.
    3. Evaluate Performance:
        - Compare mean, median, and mode.
        - Use probability to predict future performance based on historical data.
    4. Discussion and Cultural Connection:
        - Discuss how statistical analysis is used in sports management and strategy.
        - Connect to cultural preferences for different sports and their societal impact.
- Outcome: Students gain practical skills in statistics, learn to interpret data, and understand its relevance in sports and culture.

Algebra in Video Game Design

Mathematical Concept: Algebraic Expressions and Equations

Teaching Level: Secondary School (KS3/KS4)

How to Teach:

- Concept Overview: Use algebra to model in-game economies and mechanics, such as point systems or resource management.
- Cultural Context: Video games are a common interest, providing a relatable context for algebraic concepts.
- Example Problem: Model a resource management system in a game where players collect items with different values.
- Teaching Steps:
    1. Introduce the Game Scenario:
        - Explain a game where players collect coins and gems, with each having a different value.
    2. Formulate Algebraic Expressions:

- Use variables to represent unknowns (e.g., x for coins, y for gems).
- Example Equation: The total score is given by $S = 10x + 5y$.

3. Solve for Unknowns:
   - If a player needs a certain score to level up, solve for the number of coins and gems needed.
   - Example: For a score of 100, find solutions to $10x + 5y = 100$.
4. Hands-On Activity:
   - Have students create their own game scenarios using algebraic models.
   - Use coding platforms like Scratch to simulate these equations in a game.
5. Discussion:
   - Discuss the role of math in game development and its cultural impact on entertainment.
- Outcome: Students understand algebra's practical applications, appreciate its role in creative fields, and develop problem-solving skills.

Geometry in Virtual Reality and 3D Design

Mathematical Concept: 3D Geometry and Transformations

Teaching Level: Secondary School (KS3/KS4)

How to Teach:

- Concept Overview: Explore 3D shapes, transformations, and spatial reasoning, focusing on their application in virtual reality (VR) and 3D design.
- Cultural Context: The growing field of VR is exciting and relevant to many students, offering a futuristic perspective on geometry.
- Example Problem: Design a virtual room using geometric shapes and transformations.
- Teaching Steps:
    1. Introduce 3D Geometry:
       - Discuss 3D shapes (cubes, spheres, pyramids) and their properties.
    2. Explore Transformations:
       - Cover translations, rotations, and scaling of 3D objects.
       - Use software like Blender or Tinkercad to visualise these transformations.
    3. Design a Virtual Space:

- Have students design a virtual room or object, calculating dimensions and transformations.
- Example: Use transformations to create a symmetrical pattern in a virtual landscape.
  4. Cultural Element:
     - Discuss how different cultures influence architectural and design styles.
     - Explore traditional design motifs and their geometric properties.
  5. Hands-On Activity:
     - Use VR headsets or 3D software to explore and present designs.
- Outcome: Students develop spatial reasoning skills, understand the applications of geometry in technology, and appreciate cultural design influences.

Proportional Reasoning in Cooking

Mathematical Concept: Ratios and Proportional Relationships

Teaching Level: Primary and Secondary School (KS2/KS3)

How to Teach:

- Concept Overview: Use ratios and proportions to adjust recipes and understand culinary measurements.
- Cultural Context: Cooking is universal and offers opportunities to explore diverse culinary traditions.
- Example Problem: Adjust a traditional recipe to serve different numbers of people, maintaining ingredient ratios.
- Teaching Steps:
  1. Introduce Ratios:
     - Explain how ratios express relationships between quantities.
     - Example: A recipe requires 2 cups of rice for every 3 cups of water (ratio 2:3).
  2. Adjusting Recipes:
     - Task students with scaling a recipe for a family gathering.
     - Example: If the original recipe serves 4 and requires 2 cups of rice, calculate the ingredients needed for 8 servings.
     - Proportional Calculation:
$$\text{New amount of rice} = 2 \times \frac{8}{4} = 4 \, \text{cups}$$

3. Cultural Connection:
   - Explore recipes from students' cultural backgrounds, discussing unique ingredients and techniques.
4. Hands-On Activity:
   - If possible, prepare the dish in class or as a take-home assignment.
   - Encourage students to share recipes and culinary traditions from their cultures.

- Outcome: Students understand the importance of proportions in cooking, learn to apply math to everyday tasks, and explore cultural diversity through food.

Exponential Growth in Technology

Mathematical Concept: Exponential Functions and Growth

Teaching Level: Secondary School (KS4/A-Level)

How to Teach:

- Concept Overview: Study exponential functions to understand growth trends in technology, such as data storage or internet usage.
- Cultural Context: Technology is a pervasive part of modern life, impacting all students and their families.
- Example Problem: Analyse the growth of internet users over time using exponential models.
- Teaching Steps:
  1. Introduce Exponential Functions:
     - Define the general form $y = a \cdot b^x$, where a is the initial amount and b is the growth factor.

  2. Analyse Technological Growth:
     - Use historical data on internet user growth.
     - Example: In 1990, there were approximately 2.6 million users; by 2020, there were over 4 billion.
     - Model this growth: $y = 2.6 \times (1.2)^x$.
  3. Calculate Future Projections:
     - Have students calculate expected internet users in future years using the model.
     - Discuss the impact of technology on global communication and culture.

4. Hands-On Activity:
   - Use graphing software to visualise exponential growth curves.
5. Cultural Discussion:
   - Discuss how technology connects diverse cultures and the importance of understanding technological trends.

- Outcome: Students grasp exponential functions' real-world applications, appreciate technology's cultural impact, and develop analytical skills.

Calculus in Population Dynamics

Mathematical Concept: Calculus - Derivatives and Integrals

Teaching Level: Advanced Secondary School (A-Level)

How to Teach:

- Concept Overview: Introduce derivatives and integrals in the context of population dynamics, modelling how populations grow and change over time.
- Cultural Context: Different communities have unique experiences with migration and demographic changes, which can be mathematically modelled.
- Example Problem: Model the growth rate of a community's population over time, considering factors like birth and migration rates.
- Teaching Steps:
    1. Introduce Basic Calculus Concepts:
        - Define derivatives as rates of change and integrals as areas under curves.
        - Connect these concepts to real-world scenarios, such as population changes.
    2. Present the Population Model:
        - Example Equation: Use a simplified logistic growth model $\frac{dP}{dt} = rP(1 - \frac{P}{K})$, where P is the population, r is the growth rate, and K is the carrying capacity.
    3. Calculate and Interpret:
        - Differentiate the population model to find growth rates at different times.
        - Integrate to find total population change over time.
    4. Cultural Connection:
        - Discuss migration trends in students' communities and their impact on local demographics.
    5. Hands-On Activity:
        - Use graph paper or online graphing tools to visualise population models.

- Encourage students to research and present demographic changes in their communities.
- Outcome: Students understand calculus concepts in the context of real-world population dynamics, gaining analytical skills relevant to their communities.

Probability in Cultural Games

Mathematical Concept: Probability and Combinatorics

Teaching Level: Secondary School (KS3/KS4)

How to Teach:

- Concept Overview: Explore probability through traditional games from different cultures, illustrating chance and strategy.
- Cultural Context: Many cultures have traditional games that involve elements of probability and strategy, providing a rich context for exploration.
- Example Problem: Analyse a traditional African game like "Mancala" or a South Asian game like "Pachisi" to understand probabilities.
- Teaching Steps:
    1. Introduce Probability Concepts:
        - Define probability as the likelihood of an event occurring.
        - Explain basic combinatorics for calculating outcomes.
    2. Explore a Cultural Game:
        - Choose a culturally relevant game and explain its rules.
        - Example: In Mancala, calculate the probability of winning a move given certain conditions.
    3. Calculate Probabilities:
        - Analyse possible moves and their outcomes.
        - Use tree diagrams or probability tables to model outcomes.
    4. Hands-On Activity:
        - Play the game in class, encouraging students to predict and calculate probabilities for each move.
        - Discuss strategies and how probability informs decision-making.
    5. Cultural Connection:
        - Discuss the cultural significance of the game and its historical context.
- Outcome: Students learn probability through engaging, culturally relevant games, enhancing their strategic thinking and mathematical reasoning.

Functions in Mobile App Design

Mathematical Concept: Functions and Graphs

Teaching Level: Secondary School (KS4)

How to Teach:

- Concept Overview: Explore functions and their graphical representations through mobile app interfaces, which use functions to model user interactions.
- Cultural Context: Mobile technology is ubiquitous and relevant to students, offering a practical application of functions.
- Example Problem: Design a simple mobile app interface that uses linear functions to model user input and output.
- Teaching Steps:
    1. Introduce Functions:
        - Define functions as mathematical relationships between variables.
        - Explore linear functions in the form $y = mx + c$.
    2. Create a Mobile App Model:
        - Example: A budgeting app that tracks daily expenses.
        - Model daily spending with a linear function: $y = mx + c$, where m is the average daily expense and c is the initial balance.
    3. Graph the Function:
        - Use graph paper or software to plot the function.
        - Interpret the graph to predict future expenses and balances.
    4. Hands-On Activity:
        - Have students design their own app scenarios, modelling different functions.
        - Discuss real-world applications of functions in technology.
    5. Cultural Connection:
        - Explore different financial habits and needs across cultures and how technology addresses them.
- Outcome: Students understand functions through a practical application in technology, developing skills for real-world problem-solving.

Geometry in Traditional Art

Mathematical Concept: Geometry - Symmetry and Tessellation

Teaching Level: Primary and Secondary School (KS2/KS3)

How to Teach:

- Concept Overview: Study symmetry and tessellation in traditional art and design, exploring geometric patterns used in cultural artworks.
- Cultural Context: Art is a universal form of expression, with each culture having unique geometric designs and patterns.
- Example Problem: Analyse geometric patterns in Islamic tile designs or African textiles to understand symmetry and tessellation.
- Teaching Steps:
    1. Introduce Geometry Concepts:
        - Define symmetry as balanced proportions and tessellation as a repeating pattern.
        - Explore different types of symmetry (reflectional, rotational).
    2. Explore Cultural Art:
        - Study traditional patterns in different cultures, such as Moroccan tiles or African Kente cloth.
        - Identify geometric shapes and symmetries.
    3. Create and Analyse Patterns:
        - Have students create their own patterns using geometric shapes.
        - Analyse symmetry and tessellation in their designs.
    4. Hands-On Activity:
        - Use paper, rulers, and compasses to create tessellated patterns.
        - Discuss the cultural significance of different designs and their geometric properties.
    5. Cultural Connection:
        - Encourage students to research traditional art from their own cultures and share their findings.
- Outcome: Students appreciate geometry's role in art and design, understand geometric concepts, and explore cultural heritage through mathematics.

Number Theory in Cryptography

Mathematical Concept: Number Theory - Prime Numbers and Modular Arithmetic

Teaching Level: Advanced Secondary School (A-Level)

How to Teach:

- Concept Overview: Introduce number theory concepts like prime numbers and modular arithmetic, exploring their applications in cryptography.
- Cultural Context: Cryptography is crucial in digital communication, relevant to all students in an increasingly connected world.
- Example Problem: Explore the role of prime numbers in securing digital communication using the RSA encryption algorithm.
- Teaching Steps:
    1. Introduce Number Theory Concepts:
        - Define prime numbers and their properties.
        - Explain modular arithmetic and its operations.
    2. Explore Cryptography Basics:
        - Introduce basic cryptographic principles and how they secure information.
        - Example: Explain how RSA uses large prime numbers for encryption and decryption.
    3. Hands-On Activity:
        - Provide simple encryption exercises using prime numbers and modular arithmetic.
        - Have students encode and decode messages.
    4. Cultural Connection:
        - Discuss the importance of secure communication in different cultural contexts and the role of mathematics in global connectivity.
    5. Explore Ethical Considerations:
        - Discuss the ethical implications of cryptography and its impact on privacy and security.
- Outcome: Students understand number theory's application in cryptography, developing critical thinking skills and an appreciation for mathematics' role in digital security.

Complex Numbers in Engineering

Mathematical Concept: Complex Numbers

Teaching Level: Advanced Secondary School (A-Level)

How to Teach:

- Concept Overview: Explore complex numbers and their applications in engineering, focusing on electrical circuits and signal processing.
- Cultural Context: Engineering and technology are universal fields that impact students' lives and communities.
- Example Problem: Use complex numbers to model and Analyse an electrical circuit's behaviour.
- Teaching Steps:
    1. Introduce Complex Numbers:
        - Define complex numbers in the form $a + bi$, where i is the imaginary unit.
        - Explain their operations and properties.
    2. Explore Engineering Applications:
        - Introduce how complex numbers model AC circuits and signals.
        - Example: Analyse a circuit's impedance using complex numbers.
    3. Calculate and Interpret:
        - Solve problems involving complex numbers in circuit analysis.
        - Use phasor diagrams to visualise complex quantities.
    4. Hands-On Activity:
        - Use online simulations to explore circuit behaviour with complex numbers.
        - Discuss real-world applications in engineering and technology.
    5. Cultural Connection:
        - Explore the contributions of diverse cultures to engineering and technology advancements.
- Outcome: Students understand complex numbers' applications in engineering, gaining problem-solving skills and an appreciation for mathematics' role in technological development.

Global mathematics in West Africa:

1. African Counting Systems: West Africa is known for its diverse counting systems, some of which are based on groups of five or twenty. For example, the Yoruba people in Nigeria use a counting system known as "ogbo," which groups numbers in sets of twenty. Teachers can introduce these counting systems as alternative number systems, encouraging students to explore their properties and relationships to the base-10 system. Students can practice converting numbers between different systems and discover patterns within each system.

2. African Drumming and Rhythm: West African drumming traditions, such as those found in countries like Senegal and Guinea, are based on complex rhythmic patterns. These patterns

can be represented using mathematical notation, such as fractions or ratios. Teachers can introduce students to the concept of polyrhythms and explore how different drum patterns interlock. Students can analyse the ratios between the durations of different drumming parts and experiment with creating their own polyrhythms using mathematical relationships.

3. Yoruba Ife Numeration System: Equation: $N = (20 \times T) + U$. In the Ife Numeration system, the number 35 is represented as $(20 \times 1) + 15$, where $T = 1$ (one twenty) and $U = 15$ (fifteen units). The Ife Numeration system, used by the Yoruba people, combines base 20 and base 10 to represent numbers, employing geometric symbols to denote quantities.

4. Akan Adinkra symbols, such as the "Sankofa' symbol, represent concepts like learning from the past. They illustrate geometric patterns and symbolism without specific equations. Adinkra symbols are visual representations that convey meanings and messages, often based on geometric shapes and intricate patterns, representing aspects of Akan culture and wisdom.

5. Hausa Algebraic Problem-Solving: Equation: $ax + b = c$ Example: Solving the equation $2x + 3 = 9$ involves isolating the unknown variable, x, by subtracting 3 from both sides and dividing by 2. The Hausa people employed algebraic problem-solving techniques to find solutions to equations, enabling them to solve mathematical problems with unknown variables.

6. Yoruba artwork features intricate geometric patterns, showcasing symmetry, rotation, and reflection, without specific equations. Yoruba artisans incorporated geometric patterns into their art, architecture, and textiles, displaying their knowledge of geometric principles and aesthetics.

7. Hausa Geometrical Constructions: Hausa craftsmen employ geometric constructions, such as constructing perfect squares or circles, using basic tools like a compass and straightedge. Hausa craftsmen used geometric constructions to create precise shapes and structures, relying on fundamental geometric principles.

8. Gelede masks, used in Yoruba festivals, exhibit intricate patterns and designs, incorporating mathematical principles of symmetry and proportion. Explanation: Yoruba artists crafted Gelede masks with careful attention to symmetrical patterns and proportions, showcasing their understanding of mathematical concepts in their artistic expression.

9. Hausa Patterned Textiles: Hausa textiles, like "Hausa Dan Maraya" or "Ganuwar Kafi," display intricate geometric patterns and symmetrical designs. Hausa weavers and dyers incorporated mathematical principles of symmetry, repetition, and precision to create visually stunning and culturally significant textiles.

10. Yoruba Geometric Architecture: Yoruba architectural structures, like the Oba's palace in Benin City, Nigeria, incorporate geometric principles to achieve balance and aesthetic harmony. Yoruba architects and builders employed geometric principles, such as symmetry, proportion, and harmony, in the construction of palaces, shrines, and other architectural marvels.

11. Akan Geometric Symmetry in Pottery: Akan pottery demonstrates symmetrical designs and geometric patterns achieved through careful craftsmanship and attention to proportions. Akan potters utilised geometric symmetry and precise measurements to create visually appealing pottery, often adorned with intricate patterns and designs.

12. Hausa Traditional Measurement Systems: Hausa societies utilised their own measurement systems for weights, lengths, and volumes, often based on practical units like grains or seeds. Hausa communities developed measurement systems tailored to their daily needs, incorporating mathematical concepts to standardise and quantify weights, lengths, and volumes.

13. Yoruba Geometric Body Art: Yoruba body art, such as facial scarification patterns, reveals intricate geometric designs achieved through precise incisions. Yoruba body art showcases geometric patterns created through precise and deliberate scarification techniques, illustrating the application of mathematical principles to cultural practices.

14. Akan culture includes mathematical riddles and puzzles that challenge logical reasoning and problem-solving skills. Akan mathematical riddles and puzzles stimulate critical thinking and problem-solving abilities, encouraging the development of mathematical reasoning among individuals.

15. Hausa Geometric Carvings: Hausa wood carvings exhibit geometric designs and patterns, highlighting their skill in creating symmetrical and visually appealing artwork. Hausa woodcarvers incorporated geometric patterns and intricate designs into their carvings, showcasing their proficiency in utilising mathematical principles to create aesthetically pleasing art.

16. Yoruba musical instruments, like the "Dundun" drum, feature geometric patterns and proportions in their construction and ornamentation. Yoruba instrument makers employed geometric designs and proportions in crafting musical instruments, ensuring optimal sound quality and aesthetic appeal.

17. Akan masks, such as "Bwa" masks, emphasise symmetry in their designs, showcasing geometric balance and aesthetic precision. Akan mask makers meticulously crafted masks with symmetrical designs, demonstrating their understanding of geometric balance and creating visually striking cultural artifacts.

18. Hausa astronomers observed celestial phenomena, such as eclipses and solstices, employing mathematical calculations and geometric principles. Hausa astronomers used mathematical calculations and geometric principles to study celestial events, advancing their understanding of the cosmos and timekeeping.

19. Yoruba pottery displays geometric patterns and intricate designs achieved through shaping and decorative techniques. Yoruba potters applied geometric patterns and precise shaping techniques to create pottery with intricate designs, showcasing their mastery of mathematical concepts in pottery production.

20. Akan artisans incorporate fractal-like designs, featuring self-similar patterns, in their artwork. Akan artists incorporate self-similar patterns and fractal-like designs in their artwork, reflecting mathematical principles of repetition and infinite complexity.

21. Hausa culture includes mathematical proverbs that convey wisdom and knowledge through numerical concepts and mathematical metaphors. Hausa mathematical proverbs utilise numerical concepts and mathematical metaphors to convey cultural wisdom, linking mathematical thinking with everyday life.

22. Yoruba jewellery exhibits geometric shapes and patterns, such as triangles, circles, and squares, in their designs. Yoruba jewellers incorporate geometric shapes and patterns, symbolising cultural significance and expressing aesthetic appeal through their skilful craftsmanship.

23. Akan communities employed geometric principles in village planning, ensuring efficient layouts and equitable distribution of resources. Akan communities utilised geometric principles, such as symmetry and proportion, in village planning to optimise space utilisation and foster harmonious living environments.

24. Hausa Islamic architecture features intricate geometric patterns, such as tessellations and interlacing designs, inspired by Islamic artistic traditions. Hausa artisans integrated Islamic geometric patterns into their architecture, exemplifying the fusion of mathematical precision and religious symbolism.

25. Yoruba weavers incorporate geometric patterns and motifs, such as chevrons and zigzags, in their textile weaving. Yoruba weavers employ geometric patterns and motifs in their textile weaving, creating visually captivating fabrics with mathematical precision.

26. Akan drumming traditions involve complex rhythms and patterns, requiring mathematical understanding and coordination. Akan drumming relies on mathematical principles of rhythm, timing, and pattern recognition, showcasing the interplay between mathematics and music.

27. Hausa Geometric Town Layouts: Example: Hausa towns feature geometric town layouts, incorporating straight roads and orderly divisions. Explanation: Hausa town planning incorporates geometric principles, such as grid patterns and orderly divisions, to create well-structured and navigable urban spaces.

28. Yoruba body ornaments, like waist beads, showcase geometric designs and patterns, often representing cultural symbols and personal expressions. Yoruba body ornaments, adorned with geometric designs, serve as cultural adornments and personal expressions, reflecting the significance of mathematics in aesthetics.

29. Akan societies developed mathematical currency systems, employing counting methods and numerical values for trade and commerce. Akan currency systems utilised mathematical concepts, such as numerical values and counting methods, to facilitate trade and economic transactions within their communities.

30. Hausa basket weaving incorporates geometric patterns and symmetrical designs, reflecting mathematical principles in their craftsmanship. Hausa basket weavers employ geometric patterns and symmetrical designs in their weaving techniques, showcasing their proficiency in mathematical principles.

31. Yoruba dance forms, like the "Bata" dance, feature geometric body postures and movements, synchronising with rhythmic patterns. Yoruba dance traditions emphasise geometric body postures and movements, aligning with rhythmic patterns, resulting in visually captivating performances.

32. Akan puzzle games, like "Oware," involve mathematical strategies, logical reasoning, and counting skills. Akan puzzle games, such as "Oware," engage players in mathematical thinking, encouraging strategic planning and logical reasoning.

33. Hausa architecture incorporates geometric roof designs, such as pyramidal shapes and overlapping patterns. Hausa architecture features geometric roof designs, utilising shapes and patterns that provide structural stability and aesthetic appeal.

34. Yoruba hairstyles exhibit intricate geometric patterns achieved through braiding techniques and hair arrangements. Yoruba hairstylists create geometric patterns through braiding and hair arrangements, showcasing their mathematical understanding and artistic skills.

35. Akan Kente cloth displays intricate geometric patterns and colours combinations, representing cultural symbolism and historical narratives. Akan weavers incorporate geometric patterns and vibrant Colours into Kente cloth, reflecting cultural identity and aesthetic mastery.

36. Hausa pottery features geometric shapes, such as cylinders, spheres, and cones, achieved through precise pottery techniques. Hausa potters utilise mathematical principles to shape pottery into geometric forms, demonstrating their mastery of pottery craftsmanship.

37. Yoruba games, like "Ayo," involve strategic moves, counting, and mathematical thinking akin to chess. Yoruba games, including "Ayo," require players to employ strategic moves, counting skills, and mathematical thinking, fostering cognitive development and problem-solving abilities.

38. Akan woodcarvings exhibit geometric proportions and symmetry, depicting human figures and cultural motifs. Akan woodcarvers apply geometric proportions and symmetry in creating woodcarvings, emphasising cultural symbolism and aesthetic precision.

    Hausa villages incorporate geometric patterns in fencing designs, reflecting cultural identity and practicality. Hausa communities employ geometric patterns in village fencing, creating boundaries that represent cultural identity while ensuring security and protection.

39. Yoruba drum making involves precise measurements, such as drumhead size and tension, to produce desired musical tones and harmonics. Yoruba drum makers employ mathematical principles, such as measurements and tension calculations, to construct drums that produce specific musical tones and harmonics.

These examples illustrate the diverse mathematical traditions found in West Africa, spanning various disciplines, cultural practices, and artistic expressions. They highlight the deep-rooted connection between mathematics and West African societies, showcasing the ingenuity and mathematical prowess of these civilisations.

Example Lesson outline ideas and Activities:

Primary Level (Key Stage 1-2):

Lesson Title: Exploring Indigenous Number Systems

- **Objective**: To introduce students to different number systems used by Indigenous cultures around the world.

- **Activity**: Show students examples of Indigenous number systems such as Mayan, Inuit, or Aboriginal Australian. Discuss how these systems differ from the base-10 system.
- **Activity**: Have students create their own number system inspired by an Indigenous culture, using symbols or representations meaningful to them.
- **Reflection**: Ask students to reflect on the similarities and differences between their created number system and the base-10 system. Discuss the importance of diverse mathematical perspectives.

Secondary Level (Key Stage 3-4):

Lesson Title: Decolonising Geometry: Exploring Non-Euclidean Geometries

- **Objective**: To introduce students to non-Euclidean geometries and challenge Eurocentric perspectives on geometry.
- **Activity**: Present examples of non-Euclidean geometries such as spherical geometry or hyperbolic geometry. Discuss how these geometries differ from Euclidean geometry.
- **Activity**: Have students explore the properties and applications of non-Euclidean geometries through hands-on activities or interactive simulations.
- **Reflection**: Lead a class discussion on the implications of non-Euclidean geometries for our understanding of space, culture, and identity. Encourage students to critically evaluate the dominance of Euclidean geometry in the curriculum.

Post-Secondary Level (Key Stage 5 and Higher Education):

Lesson Title: Decolonising Calculus: Exploring Indigenous Mathematical Practices

- **Objective**: To examine Indigenous mathematical practices and their relevance to calculus.
- **Activity**: Introduce students to examples of Indigenous mathematical practices such as counting systems, geometric patterns, or navigational techniques.
- **Activity**: Have students investigate how Indigenous mathematical knowledge intersects with concepts in calculus, such as rates of change or geometric transformations.
- **Reflection**: Facilitate a seminar discussion on the role of Indigenous knowledge in mathematics and the importance of incorporating diverse perspectives into higher education curriculum.

Primary Level (Key Stage 1-2):

Lesson Title: Exploring Indigenous Counting Systems

- **Key Concept**: Counting and Place Value
- **Lesson Length**: 1 hour

- **Objective**: To introduce students to Indigenous counting systems and develop understanding of place value.
- **Activity**: Present examples of Indigenous counting systems such as the Mayan vigesimal system or Aboriginal Australian counting sticks. Discuss the concepts of base-10 and place value.
- **Activity**: Have students create their own counting system inspired by an Indigenous culture, using symbols or representations meaningful to them. Practice counting using their created system.
- **Assessment**: Students demonstrate their understanding by accurately counting objects using both the base-10 system and their created Indigenous counting system.

Secondary Level (Key Stage 3-4):

Lesson Title: Decolonising Geometry: Exploring Non-Euclidean Shapes

- **Key Concept**: Geometry and Shape
- **Lesson Length**: 1.5 hours
- **Objective**: To introduce students to non-Euclidean geometries and challenge Eurocentric perspectives on geometry.
- **Activity**: Present examples of non-Euclidean shapes such as the sphere, hyperbolic plane, or fractals. Discuss how these shapes differ from Euclidean shapes and their applications.
- **Activity**: Have students explore the properties of non-Euclidean shapes through interactive simulations or hands-on activities. Compare and contrast their properties with Euclidean shapes.
- **Assessment**: Students demonstrate their understanding by identifying and describing the properties of non-Euclidean shapes and their significance in real-world contexts.

Post-Secondary Level (Key Stage 5 and Higher Education):

Lesson Title: Decolonising Calculus: Exploring Indigenous Mathematical Practices

- **Key Concept**: Calculus and Mathematical Practices
- **Lesson Length**: 2 hours
- **Objective**: To examine Indigenous mathematical practices and their relevance to calculus.
- **Activity**: Introduce students to examples of Indigenous mathematical practices such as counting systems, geometric patterns, or navigational techniques.
- **Activity**: Have students investigate how Indigenous mathematical knowledge intersects with concepts in calculus, such as rates of change or integration. Explore applications of Indigenous mathematics in fields such as astronomy, agriculture, or architecture.
- **Assessment**: Students demonstrate their understanding by presenting a research project or essay exploring the connections between Indigenous mathematical practices and calculus concepts, providing examples and evidence to support their analysis.

Primary Level (Key Stage 1-2):

Lesson Title: Cultural Patterns in Shape and Symmetry

- **Key Concept**: Geometry and Symmetry
- **Lesson Length**: 1 hour
- **Objective**: To explore cultural patterns and symmetry in shapes.
- **Activity**: Introduce students to patterns and designs from different cultures, such as Islamic geometric patterns, African textiles, or Indigenous artworks.
- **Activity**: Have students create their own symmetrical designs inspired by patterns from various cultures, using shapes and colours. Discuss the concepts of reflectional and rotational symmetry.
- **Assessment**: Students demonstrate their understanding by explaining the symmetrical properties of their designs and identifying cultural influences in their patterns.

Secondary Level (Key Stage 3-4):

Lesson Title: Decolonising Algebra: Exploring Algebraic Thinking in Different Cultures

- **Key Concept**: Algebraic Thinking
- **Lesson Length**: 1.5 hours
- **Objective**: To investigate algebraic thinking in different cultural contexts.
- **Activity**: Present examples of algebraic thinking from diverse cultures, such as ancient Babylonian algebra or Chinese algebraic methods.
- **Activity**: Have students explore algebraic problem-solving techniques used in different cultures, such as the use of geometric diagrams or verbal reasoning.
- **Assessment**: Students demonstrate their understanding by solving algebraic problems inspired by different cultural contexts and explaining their solution methods.

Post-Secondary Level (Key Stage 5 and Higher Education):

Lesson Title: Indigenous Mathematical Models in Probability and Statistics

- **Key Concept**: Probability and Statistics
- **Lesson Length**: 2 hours
- **Objective**: To examine Indigenous mathematical models in probability and statistics.
- **Activity**: Introduce students to Indigenous approaches to probability and statistics, such as traditional methods of weather prediction or land management practices.

- **Activity**: Have students analyse data sets related to Indigenous communities or environmental issues using Indigenous mathematical models. Discuss the strengths and limitations of these models compared to Western statistical methods.
- **Assessment**: Students demonstrate their understanding by applying Indigenous mathematical models to analyse real-world data sets and presenting their findings in a written report or presentation.

These lesson plans offer opportunities for students to engage with mathematics in culturally relevant and meaningful ways, promoting diversity and inclusivity in the mathematics classroom.

Primary Level (Key Stage 1-2):

Lesson Title: Traditional Measurement Systems

- **Key Concept**: Measurement
- **Lesson Length**: 1 hour
- **Objective**: To explore traditional measurement systems from different cultures.
- **Activity**: Introduce students to traditional measurement systems such as cubits, hand spans, or fathoms used by ancient civilisations or indigenous communities.
- **Activity**: Have students engage in hands-on measurement activities using traditional units of measurement. For example, measure the length of objects using hand spans or the area of a surface using cubits.
- **Assessment**: Students demonstrate their understanding by comparing and converting between traditional units of measurement and standard metric units.

Secondary Level (Key Stage 3-4):

Lesson Title: Cultural Perspectives on Data Representation

- **Key Concept**: Data Handling and Representation
- **Lesson Length**: 1.5 hours
- **Objective**: To examine cultural perspectives on data representation and visualisation.
- **Activity**: Present examples of data visualisation techniques from different cultures, such as Inca quipus, Aboriginal Australian songlines, or Maori carvings.
- **Activity**: Have students create their own data visualisations inspired by cultural practices, using symbols, patterns, or storytelling techniques to represent data.
- **Assessment**: Students demonstrate their understanding by explaining the cultural significance of their data visualisations and interpreting data presented in different cultural contexts.

Post-Secondary Level (Key Stage 5 and Higher Education):

Lesson Title: Indigenous Mathematics and Environmental Sustainability

- **Key Concept**: Applied Mathematics and Environmental Science
- **Lesson Length**: 2 hours
- **Objective**: To explore Indigenous mathematical principles and their application to environmental sustainability.
- **Activity**: Introduce students to Indigenous concepts of environmental stewardship and sustainability, such as the Indigenous Knowledge Systems framework or the concept of reciprocity in Indigenous cultures.
- **Activity**: Have students analyse environmental data using Indigenous mathematical models and principles, such as traditional ecological knowledge or sustainable resource management practices.
- **Assessment**: Students demonstrate their understanding by proposing solutions to environmental challenges based on Indigenous mathematical principles and presenting their findings in a group discussion or written report.

Further Education Level (Higher Education and Beyond):

Lesson Title: Decolonising Mathematics Education: Critical Perspectives

- **Key Concept**: Critical Mathematics Education
- **Lesson Length**: 2.5 hours
- **Objective**: To critically examine the Eurocentric biases in mathematics education and explore decolonising approaches.
- **Activity**: Facilitate a seminar discussion on the historical and cultural roots of Eurocentrism in mathematics education, focusing on colonialism, imperialism, and cultural hegemony.
- **Activity**: Have students analyse curriculum materials, textbooks, and assessments for Eurocentric biases and propose alternative perspectives and content that reflect diverse cultural experiences and knowledge systems.
- **Assessment**: Students demonstrate their understanding by writing a reflective essay or engaging in a debate on the implications of decolonising mathematics education for teaching, learning, and social justice.

Primary Level (Key Stage 1-2):

Lesson Title: Cultural Patterns in Data Representation

- **Key Concept**: Data Handling and Representation
- **Lesson Length**: 1 hour
- **Objective**: To explore cultural patterns in data representation and interpretation.

- **Activity**: Introduce students to different cultural symbols and patterns used in data representation, such as pictograms, ideograms, or traditional designs.
- **Activity**: Have students create their own cultural data representations using symbols or patterns from their own cultural backgrounds, and share them with the class.
- **Assessment**: Students demonstrate their understanding by explaining the cultural significance of their data representations and interpreting data presented in different cultural contexts.

Secondary Level (Key Stage 3-4):

Lesson Title: Decolonising Algebra: Exploring Algebraic Thinking in Ancient Civilisations

- **Key Concept**: Algebraic Thinking
- **Lesson Length**: 1.5 hours
- **Objective**: To investigate algebraic thinking in ancient civilisations and its relevance to modern mathematics.
- **Activity**: Present examples of algebraic problem-solving techniques used in ancient civilisations such as Mesopotamia, Egypt, or China.
- **Activity**: Have students solve algebraic problems inspired by ancient civilisations, using methods and techniques from historical sources.
- **Assessment**: Students demonstrate their understanding by solving algebraic problems using both modern and ancient algebraic methods, and comparing their effectiveness.

Post-Secondary Level (Key Stage 5 and Higher Education):

Lesson Title: Indigenous Mathematics and Social Justice

- **Key Concept**: Applied Mathematics and Social Justice
- **Lesson Length**: 2 hours
- **Objective**: To explore the intersection of Indigenous mathematics and social justice issues.
- **Activity**: Introduce students to Indigenous mathematical concepts such as circularity, reciprocity, or relationality, and discuss their implications for social justice.
- **Activity**: Have students analyse social justice data using Indigenous mathematical models and principles, and propose solutions to address systemic inequalities and injustices.
- **Assessment**: Students demonstrate their understanding by presenting a research project or policy proposal that applies Indigenous mathematical principles to real-world social justice issues.

Further Education Level (Higher Education and Beyond):

Lesson Title: Decolonising Mathematics Education: Global Perspectives

- **Key Concept:** Critical Mathematics Education
- **Lesson Length:** 2.5 hours
- **Objective:** To critically examine global perspectives on decolonising mathematics education.
- **Activity:** Facilitate a seminar discussion on decolonising initiatives and challenges in different cultural contexts around the world, such as Africa, Asia, or the Americas.
- **Activity:** Have students compare and contrast decolonising approaches and strategies from different cultural perspectives, and reflect on the implications for their own teaching practice.
- **Assessment:** Students demonstrate their understanding by writing a comparative analysis or engaging in a group presentation on global perspectives on decolonising mathematics education.

Primary Level (Key Stage 1-2):

Lesson Title: Exploring Indigenous Geometry: Understanding Shape and Space

- **Key Concept:** Geometry and Spatial Reasoning
- **Lesson Length:** 1 hour
- **Objective:** To introduce students to Indigenous geometric shapes and their cultural significance.
- **Activity:** Present examples of Indigenous geometric shapes such as mandalas, dreamcatchers, or Maori taonga pūoro (carvings). Discuss the cultural meanings and symbolism behind these shapes.
- **Activity:** Have students create their own Indigenous-inspired geometric shapes using natural materials or art supplies. Encourage them to explore concepts of symmetry, pattern, and proportion.
- **Assessment:** Students demonstrate their understanding by explaining the mathematical properties of their geometric shapes and their cultural significance.

Secondary Level (Key Stage 3-4):

Lesson Title: Decolonising Algebra: Exploring Algebraic Patterns in Nature

- **Key Concept:** Algebraic Patterns and Functions
- **Lesson Length:** 1.5 hours
- **Objective:** To investigate algebraic patterns in natural phenomena and cultural artifacts.
- **Activity:** Present examples of algebraic patterns found in nature, such as Fibonacci sequences in sunflowers, or fractal patterns in coastlines. Discuss the mathematical principles behind these patterns.

- **Activity**: Have students analyse cultural artifacts such as traditional textiles, pottery, or architecture to identify algebraic patterns and symmetries. Explore the connections between mathematics, art, and culture.
- **Assessment**: Students demonstrate their understanding by identifying and describing algebraic patterns in natural phenomena and cultural artifacts, and explaining their significance.

Post-Secondary Level (Key Stage 5 and Higher Education):

Lesson Title: Indigenous Calculus: Understanding Rates of Change in Traditional Knowledge Systems

- **Key Concept**: Calculus and Rates of Change
- **Lesson Length**: 2 hours
- **Objective**: To explore Indigenous approaches to understanding rates of change and dynamic systems.
- **Activity**: Introduce students to Indigenous concepts of time, change, and cyclical patterns, such as seasonal calendars or lunar cycles. Discuss how Indigenous knowledge systems incorporate concepts of calculus-like reasoning.
- **Activity**: Have students analyse traditional ecological knowledge systems or Indigenous agricultural practices to identify examples of rates of change and dynamic systems. Explore the mathematical principles underlying these practices.
- **Assessment**: Students demonstrate their understanding by applying calculus concepts to analyse Indigenous knowledge systems and explain the connections between mathematics, culture, and sustainability.

Further Education Level (Higher Education and Beyond):

Lesson Title: Decolonising Probability and Statistics: Exploring Cultural Data Sets

- **Key Concept**: Probability and Statistics
- **Lesson Length**: 2.5 hours
- **Objective**: To critically examine cultural perspectives on probability and statistics.
- **Activity**: Present examples of cultural data sets and statistical analyses from different cultural contexts, such as health disparities in Indigenous communities or economic inequalities in post-colonial societies.
- **Activity**: Have students conduct statistical analyses of cultural data sets, considering the social, historical, and cultural factors that influence the data. Discuss the ethical implications of statistical research in diverse cultural contexts.

- **Assessment**: Students demonstrate their understanding by conducting a statistical analysis of a cultural data set, interpreting the results in relation to cultural factors, and reflecting on the ethical considerations involved.

These lesson plans provide opportunities for students to engage with difficult-to-explain mathematical concepts while also exploring cultural perspectives and relevance. By connecting mathematics to culture, students gain a deeper understanding of abstract mathematical concepts and their real-world applications.

Primary Level (Key Stage 1-2):

Lesson Title: Caribbean Counting Rhymes: Exploring Number Patterns

- **Key Concept**: Number Patterns and Counting
- **Lesson Length**: 1 hour
- **Objective**: To introduce students to Caribbean counting rhymes and patterns.
- **Activity**: Introduce students to traditional Caribbean counting rhymes such as "One, Two, Buckle My Shoe" or "Miss Mary Mack." Discuss the rhythmic patterns and sequences present in these rhymes.
- **Activity**: Have students create their own counting rhymes inspired by Caribbean culture, incorporating cultural symbols, language, and rhythms. Practice counting and identifying number patterns in the rhymes.
- **Assessment**: Students demonstrate their understanding by reciting their counting rhymes, identifying number patterns, and explaining the cultural significance of their rhymes.

Secondary Level (Key Stage 3-4):

Lesson Title: African Fractals: Exploring Geometric Patterns

- **Key Concept**: Geometry and Fractal Patterns
- **Lesson Length**: 1.5 hours
- **Objective**: To explore African fractals and their mathematical properties.
- **Activity**: Introduce students to examples of African fractals such as traditional African art, architecture, or textiles. Discuss the self-similarity and recursive patterns found in these fractals.
- **Activity**: Have students create their own fractal designs inspired by African patterns, using geometric shapes and iterative processes. Explore the mathematical principles behind fractal geometry.
- **Assessment**: Students demonstrate their understanding by creating and analysing their fractal designs, identifying mathematical properties such as self-similarity and scale invariance.

Post-Secondary Level (Key Stage 5 and Higher Education):

Lesson Title: South Asian Vedic Mathematics: Exploring Mental Calculation Techniques

- **Key Concept**: Arithmetic and Mental Calculation
- **Lesson Length**: 2 hours
- **Objective**: To investigate Vedic mathematics techniques from South Asia and their application to mental calculation.
- **Activity**: Introduce students to Vedic mathematics techniques such as sutras (mathematical formulas) and vedic mathematics tricks for mental calculation. Discuss the historical and cultural context of Vedic mathematics.
- **Activity**: Have students practice mental calculation using Vedic mathematics techniques, such as the sutra for multiplication or the concept of digital roots. Explore the efficiency and versatility of these techniques.
- **Assessment**: Students demonstrate their understanding by solving arithmetic problems using Vedic mathematics techniques and comparing their efficiency to standard methods.

Further Education Level (Higher Education and Beyond):

Lesson Title: Caribbean Probability and Games: Exploring Cultural Probability

- **Key Concept**: Probability and Games of Chance
- **Lesson Length**: 2.5 hours
- **Objective**: To examine Caribbean cultural perspectives on probability and games of chance.
- **Activity**: Introduce students to traditional Caribbean games of chance such as "All Fours" or "Ludi." Discuss the mathematical principles and probabilities involved in these games.
- **Activity**: Have students analyse the probabilities of different outcomes in Caribbean games of chance, considering factors such as strategy, luck, and cultural norms. Discuss the social and cultural significance of these games.
- **Assessment**: Students demonstrate their understanding by calculating probabilities and expected outcomes in Caribbean games of chance, and reflecting on the cultural factors that influence gameplay.

Primary Level (Key Stage 1-2):

Lesson Title: Caribbean Storytelling Mathematics: Exploring Numerical Patterns

- **Key Concept**: Numerical Patterns and Storytelling
- **Lesson Length**: 1 hour
- **Objective**: To introduce students to Caribbean storytelling traditions and their connection to numerical patterns.

- **Activity**: Share traditional Caribbean folktales or stories that incorporate numerical patterns, such as Anansi stories or tales of Caribbean folklore heroes. Discuss the rhythmic and repetitive elements in these stories.
- **Activity**: Have students identify numerical patterns in the stories, such as counting sequences, repetitive phrases, or numerical symbolism. Create visual representations of the numerical patterns using illustrations or charts.
- **Assessment**: Students demonstrate their understanding by retelling the stories, identifying numerical patterns, and explaining the cultural significance of the stories.

Secondary Level (Key Stage 3-4):

Lesson Title: Caribbean Calypso Mathematics: Exploring Algebraic Expression

- **Key Concept**: Algebraic Expression and Symbolism
- **Lesson Length**: 1.5 hours
- **Objective**: To explore algebraic expression through Caribbean calypso music.
- **Activity**: Introduce students to Caribbean calypso music and its use of metaphor, symbolism, and wordplay. Analyse calypso lyrics for mathematical expressions, equations, or patterns.
- **Activity**: Have students create their own calypso-inspired mathematical expressions or equations, using Caribbean themes, language, and cultural references. Write and perform their mathematical calypso songs in groups.
- **Assessment**: Students demonstrate their understanding by explaining the mathematical expressions in their calypso songs, identifying mathematical patterns and symbolism, and performing their songs for the class.

Post-Secondary Level (Key Stage 5 and Higher Education):

Lesson Title: Caribbean Carnival Statistics: Exploring Probability and Data Analysis

- **Key Concept**: Probability and Data Analysis
- **Lesson Length**: 2 hours
- **Objective**: To investigate probability and data analysis through Caribbean carnival traditions.
- **Activity**: Introduce students to Caribbean carnival traditions such as masquerade bands, costume design, and parade routes. Discuss the statistical concepts involved in carnival planning and execution.
- **Activity**: Have students collect and analyse carnival-related data such as costume design preferences, parade attendance figures, or festival expenditure. Use statistical techniques to analyse the data and draw conclusions about carnival trends and patterns.
- **Assessment**: Students demonstrate their understanding by presenting their data analysis findings, interpreting statistical results, and proposing recommendations for improving carnival planning and management.

Further Education Level (Higher Education and Beyond):

Lesson Title: Caribbean Cultural Mathematics: Exploring Ethnomathematics

- **Key Concept:** Ethnomathematics and Cultural Mathematics
- **Lesson Length:** 2.5 hours
- **Objective:** To examine Caribbean cultural perspectives on mathematics and explore ethnomathematical concepts.
- **Activity:** Introduce students to ethnomathematics and its relevance to Caribbean cultures, including concepts such as counting systems, geometric designs, and mathematical rituals.
- **Activity:** Have students conduct research on specific aspects of Caribbean ethnomathematics, such as traditional games, folklore, or spiritual practices that involve mathematical concepts. Present their findings to the class and discuss the cultural significance of these mathematical traditions.
- **Assessment:** Students demonstrate their understanding by writing a reflective essay or research paper on Caribbean ethnomathematics, analysing specific examples and exploring the implications for mathematics education and cultural identity.

Primary Level (Key Stage 1-2):

Lesson Title: Ancient African Number Systems: Exploring Numerical Representation

- **Key Concept:** Numerical Representation and Counting Systems
- **Lesson Length:** 1 hour
- **Objective:** To introduce students to ancient African number systems and their unique numerical representation.
- **Activity:** Introduce students to ancient African number systems such as the Egyptian hieroglyphic numerals or the Nsibidi script from West Africa. Discuss the symbols used for counting and numerical representation.
- **Activity:** Have students create their own numerical symbols inspired by ancient African number systems, using geometric shapes and patterns. Practice counting and representing numbers using their created symbols.
- **Assessment:** Students demonstrate their understanding by explaining the numerical symbols they created and using them to represent and count numbers.

Secondary Level (Key Stage 3-4):

Lesson Title: African Geometric Patterns: Exploring Symmetry and Design

- **Key Concept:** Geometry and Symmetry
- **Lesson Length:** 1.5 hours

- **Objective**: To explore geometric patterns from ancient African cultures and their mathematical properties.
- **Activity**: Introduce students to examples of geometric patterns from ancient African civilisations such as the fractal designs of the Benin bronzes or the intricate textile patterns of the Kente cloth from Ghana. Discuss the symmetrical properties and mathematical principles behind these patterns.
- **Activity**: Have students create their own geometric patterns inspired by ancient African designs, using rotational and reflectional symmetry. Explore the mathematical concepts of tessellation and repetition.
- **Assessment**: Students demonstrate their understanding by explaining the symmetrical properties of their geometric patterns and identifying mathematical principles present in ancient African designs.

Post-Secondary Level (Key Stage 5 and Higher Education):

Lesson Title: African Mathematical Philosophy: Exploring Mathematical Thought

- **Key Concept**: Mathematical Philosophy and Ethnomathematics
- **Lesson Length**: 2 hours
- **Objective**: To examine the philosophical foundations of mathematics in ancient African cultures.
- **Activity**: Introduce students to the mathematical philosophy of ancient African civilisations such as the Yoruba of Nigeria or the Egyptians. Discuss concepts such as interconnectedness, harmony, and balance in mathematical thought.
- **Activity**: Have students analyse historical texts or artifacts related to African mathematical philosophy, such as the Rhind Mathematical Papyrus or the Ishango bone. Discuss the mathematical principles and cultural significance of these artifacts.
- **Assessment**: Students demonstrate their understanding by writing a reflective essay or research paper on African mathematical philosophy, exploring its implications for modern mathematics and society.

Further Education Level (Higher Education and Beyond):

Lesson Title: African Ethnomathematics: Exploring Mathematical Practices

- **Key Concept**: Ethnomathematics and Cultural Mathematics
- **Lesson Length**: 2.5 hours
- **Objective**: To investigate ethnomathematical concepts and practices from ancient African cultures.
- **Activity**: Introduce students to specific examples of ethnomathematical practices from ancient African civilisations, such as geometric construction techniques, calendar systems, or

navigation methods. Discuss the mathematical principles and cultural contexts of these practices.

- **Activity**: Have students conduct research on a chosen aspect of African ethnomathematics, such as the mathematical principles behind African architecture or the mathematical symbolism in African art. Present their findings to the class and lead a discussion on the cultural significance of these mathematical traditions.
- **Assessment**: Students demonstrate their understanding by presenting their research findings, analysing specific examples of African ethnomathematics, and reflecting on the implications for mathematics education and cultural identity.

Primary Level (Key Stage 1-2):

Lesson Title: Ancient African Algebra: Exploring Algebraic Patterns

- **Key Concept**: Algebraic Patterns and Expressions
- **Lesson Length**: 1 hour
- **Objective**: To introduce students to algebraic patterns inspired by ancient African cultures.
- **Activity**: Introduce students to examples of algebraic patterns found in ancient African artifacts or designs, such as repeating geometric motifs or rhythmic drumming patterns. Discuss the concepts of variables, constants, and patterns.
- **Activity**: Have students create their own algebraic patterns inspired by ancient African designs, using symbols or shapes to represent variables and constants. Practice identifying and extending patterns in their algebraic expressions.
- **Assessment**: Students demonstrate their understanding by explaining the patterns in their algebraic expressions and identifying the variables and constants used.

Secondary Level (Key Stage 3-4):

Lesson Title: African Geometric Constructions: Exploring Circumference and Pi

- **Key Concept**: Geometry and Circumference
- **Lesson Length**: 1.5 hours
- **Objective**: To explore geometric constructions and the concept of circumference inspired by ancient African methods.
- **Activity**: Introduce students to ancient African methods of geometric construction, such as the use of ropes or knotted strings for measuring circumferences. Discuss the concept of pi and its relationship to circumference.
- **Activity**: Have students experiment with geometric constructions to measure circumferences using African-inspired methods, such as using ropes or strings to create circles and estimate their circumferences. Compare their measurements with the formula for circumference.

- **Assessment**: Students demonstrate their understanding by calculating circumferences using both African-inspired methods and the circumference formula, and comparing their results.

Post-Secondary Level (Key Stage 5 and Higher Education):

Lesson Title: Ancient African Algebraic Equations: Exploring Equivalence and Balance

- **Key Concept**: Algebraic Equations and Equivalence
- **Lesson Length**: 2 hours
- **Objective**: To investigate ancient African concepts of algebraic equations and balance.
- **Activity**: Introduce students to examples of algebraic equations and problem-solving techniques from ancient African civilisations, such as the use of scales or balance beams for solving equations. Discuss the concept of equivalence and balancing equations.
- **Activity**: Have students solve algebraic equations inspired by ancient African methods, using balance and symmetry as problem-solving tools. Practice solving equations by manipulating terms to maintain balance.
- **Assessment**: Students demonstrate their understanding by solving algebraic equations using both traditional algebraic methods and ancient African-inspired techniques, and explaining the principles of balance and equivalence.

Further Education Level (Higher Education and Beyond):

Lesson Title: African Mathematical Astronomy: Exploring Angular Measurement

- **Key Concept**: Trigonometry and Angular Measurement
- **Lesson Length**: 2.5 hours
- **Objective**: To examine ancient African methods of angular measurement and their applications in mathematical astronomy.
- **Activity**: Introduce students to ancient African methods of angular measurement, such as using sundials or star charts for tracking celestial movements. Discuss the mathematical principles of angular measurement and trigonometry.
- **Activity**: Have students analyse ancient African astronomical artifacts or texts, such as the Zimbabwean Great Zimbabwe ruins or Egyptian star charts. Explore the mathematical techniques used for measuring angles and predicting celestial events.
- **Assessment**: Students demonstrate their understanding by interpreting ancient African astronomical artifacts or texts, calculating angular measurements, and explaining the mathematical principles behind ancient African methods of angular measurement.

Primary Level (Key Stage 1-2):

Lesson Title: African and Caribbean Number Stories

- **Key Concept**: Developing fluency in number and place value.
- **Lesson Length**: 1 hour
- **Objective**: To consolidate numerical capability by solving number stories inspired by African and Caribbean cultures.
- **Activity**: Introduce students to traditional African and Caribbean number stories that involve counting, addition, and subtraction. These stories can feature cultural elements like animals, folklore characters, or traditional activities.
- **Activity**: Have students create their own number stories inspired by African and Caribbean culture. Encourage them to incorporate traditional characters or settings from stories they've heard.
- **Assessment**: Students demonstrate fluency by solving the number stories they created and explaining the cultural elements included.

Secondary Level (Key Stage 3-4):

Lesson Title: African and Caribbean Algebraic Patterns

- **Key Concept**: Using algebra to generalise arithmetic and solve equations.
- **Lesson Length**: 1.5 hours
- **Objective**: To use algebraic patterns inspired by African and Caribbean cultures to solve equations.
- **Activity**: Introduce students to algebraic patterns found in African and Caribbean art, music, or storytelling traditions. Discuss how patterns are used to represent relationships and solve problems.
- **Activity**: Have students create their own algebraic patterns inspired by African and Caribbean cultural motifs. They should then write equations to describe and solve these patterns.
- **Assessment**: Students demonstrate their algebraic fluency by solving equations based on the patterns they created and explaining the cultural significance of their patterns.

Post-Secondary Level (Key Stage 5 and Higher Education):

Lesson Title: African and Caribbean Geometric Constructions

- **Key Concept**: Understanding geometric constructions and applying them to real-world contexts.
- **Lesson Length**: 2 hours
- **Objective**: To explore geometric constructions used in African and Caribbean architecture and art.

- **Activity**: Introduce students to geometric constructions found in African and Caribbean architecture, such as the use of symmetry and tessellation in traditional buildings and artwork.
- **Activity**: Have students analyse geometric constructions in African and Caribbean art and architecture. They should then create their own geometric designs inspired by these traditions, using compass and ruler constructions.
- **Assessment**: Students demonstrate their understanding by constructing geometric designs based on African and Caribbean cultural motifs and explaining the mathematical principles behind their designs.

Further Education Level (Higher Education and Beyond):

Lesson Title: African and Caribbean Statistical Investigations

- **Key Concept**: Conducting statistical investigations using real-world data.
- **Lesson Length**: 2.5 hours
- **Objective**: To explore statistical concepts using data from African and Caribbean contexts.
- **Activity**: Provide students with datasets related to African and Caribbean demographics, economics, or social indicators. Guide them through analysing the data using statistical techniques such as mean, median, mode, and range.
- **Activity**: Have students conduct their own statistical investigations using data from African and Caribbean sources. They should then present their findings and discuss the cultural and societal implications of their results.
- **Assessment**: Students demonstrate their statistical reasoning skills by conducting and presenting their own investigations using African and Caribbean data sets.

Primary Level (Key Stage 1-2):

Lesson Title: African and Caribbean Number Rhymes

- **Key Concept**: Developing fluency in number and place value.
- **Lesson Length**: 1 hour
- **Objective**: To consolidate numerical capability through rhythmic number rhymes inspired by African and Caribbean cultures.
- **Activity**: Introduce students to traditional African and Caribbean number rhymes that involve counting, sequencing, and basic arithmetic operations.
- **Activity**: Have students create their own number rhymes inspired by African and Caribbean cultural elements such as traditional instruments, dance movements, or local animals.
- **Assessment**: Students demonstrate fluency by performing their number rhymes and explaining the cultural significance of the elements included.

Secondary Level (Key Stage 3-4):

Lesson Title: African and Caribbean Algebraic Expressions

- **Key Concept**: Using algebraic expressions to solve problems.
- **Lesson Length**: 1.5 hours
- **Objective**: To use algebraic expressions inspired by African and Caribbean patterns to solve equations.
- **Activity**: Introduce students to algebraic expressions found in African and Caribbean art, music, or cultural traditions. Discuss how patterns and symbolism are used to represent mathematical relationships.
- **Activity**: Have students analyse algebraic expressions in African and Caribbean artifacts or cultural practices. They should then create their own algebraic expressions inspired by these traditions, incorporating symbols and patterns.
- **Assessment**: Students demonstrate their algebraic fluency by solving equations based on the expressions they created and explaining the cultural significance of their patterns.

Post-Secondary Level (Key Stage 5 and Higher Education):

Lesson Title: African and Caribbean Geometry in Architecture

- **Key Concept**: Understanding geometric principles in architectural design.
- **Lesson Length**: 2 hours
- **Objective**: To explore geometric constructions used in African and Caribbean architecture and apply them to real-world design challenges.
- **Activity**: Introduce students to geometric principles found in African and Caribbean architecture, such as symmetry, tessellation, and fractal patterns. Discuss how these principles are used to create visually stunning and structurally sound buildings.
- **Activity**: Have students analyse architectural designs from Africa and the Caribbean, focusing on geometric elements and cultural symbolism. They should then create their own architectural designs inspired by these traditions, incorporating geometric principles and cultural motifs.
- **Assessment**: Students demonstrate their understanding by presenting their architectural designs and explaining the mathematical and cultural considerations behind their creations.

Further Education Level (Higher Education and Beyond):

Lesson Title: Statistical Analysis of African and Caribbean Demographics

- **Key Concept**: Conducting statistical analyses using real-world data.
- **Lesson Length**: 2.5 hours

- **Objective**: To explore statistical concepts using demographic data from African and Caribbean countries.
- **Activity**: Provide students with demographic data sets related to African and Caribbean populations, including variables such as age, gender, education, and employment. Guide them through analysing the data using statistical techniques such as descriptive statistics, correlation analysis, and regression modelling.
- **Activity**: Have students conduct their own statistical analyses using demographic data from African and Caribbean countries. They should then present their findings and discuss the cultural, social, and economic implications of their results.
- **Assessment**: Students demonstrate their statistical reasoning skills by conducting and presenting their own analyses using African and Caribbean demographic data sets.

Activities and Projects Promoting Inclusive Mathematics Learning:

1. **Cultural Mathematics Journals**:
    - Encourage students to keep mathematics journals where they explore how mathematical concepts are applied in different cultural contexts. They can research and write about traditional mathematical practices from various cultures and reflect on the connections between culture and mathematics.

2. **Mathematical Storytelling**:
    - Have students create mathematical stories or narratives that incorporate diverse characters, settings, and cultural elements. This activity promotes creativity, literacy, and mathematical reasoning while celebrating cultural diversity.

3. **Community Mathematics Projects**:
    - Collaborate with local community organisations or cultural institutions to develop mathematics projects that address real-world issues facing diverse communities. Students can use mathematical concepts to analyse data, solve problems, and propose solutions that are relevant and meaningful to their community.

4. **Mathematics of Art and Architecture**:
    - Explore the mathematical principles underlying art and architecture from different cultural traditions. Students can study geometric patterns in Islamic art, fractal geometry in African designs, or symmetry in Indigenous motifs, fostering appreciation for cultural diversity and mathematical beauty.

5. **Global Mathematics Challenges**:
    - Participate in international mathematics challenges or competitions that feature problems inspired by diverse cultural contexts. Students can work collaboratively to

solve problems that draw on mathematical concepts from around the world, promoting cross-cultural exchange and learning.

Signposts for diversity in mathematics

West African mathematical traditions to encourage exploration, problem-solving, and critical thinking:

1. Explore the rich symbolism of Adinkra symbols from West Africa and create your own designs using geometric shapes and patterns. Research the meanings behind the symbols and discuss their mathematical representations.

2. Fractal Drumming: Investigate the mathematical patterns and rhythms found in West African drumming. Learn traditional drumming techniques and explore the self-similarity and repetitive patterns in the music.

3. Create a set of pattern tiles inspired by African textile designs. Use symmetry, rotation, and tessellation to design unique patterns. Discuss the mathematical concepts behind the designs.

4. Design a mase inspired by Ananse, the mythical spider character from West African folklore. Use logic and problem-solving skills to create challenging paths and obstacles within the mase.

5. Learn traditional African bead weaving techniques and create intricate patterns using different colours and sizes of beads. Explore mathematical concepts such as symmetry, sequences, and patterns in your designs.

6. Investigate the Ashanti number system and compare it to the decimal system. Create your own number system based on symbols and explore the mathematical operations using this system.

7. Explore Sona, the traditional art of geometric sand drawing from the Dogon people of Mali. Use compasses, rulers, and coloured sand to create geometric patterns and discuss the mathematical principles behind them.

8. Play the traditional African game of Mancala and analyse different strategies and tactics. Use critical thinking skills to develop your own winning strategies and explore mathematical concepts such as counting, probability, and strategic thinking.

9. Design and construct geometric masks inspired by African art. Explore symmetry, shape, and proportion while incorporating cultural elements into your creations.

10. Learn about the mathematical principles behind constructing a Djembe drum. Explore measurements, angles, and ratios while building your own drum or studying the designs of existing drums.

11. Experiment with traditional Nigerian Adire textile dyeing techniques and create geometric patterns using resist dyeing methods. Discuss the mathematical properties of the patterns created.

12. Solve geometric puzzles inspired by African designs. Explore concepts such as tangrams, polyominoes, and dissection puzzles while discovering the mathematical patterns within the puzzles.

13. Study the Nsibidi writing system from Nigeria and create your own coded messages using the symbols. Decode messages created by others and discuss the mathematical relationships between symbols.

14. Research and create models of iconic African architecture, such as the Great Mosque of Djenné or the Pyramids of Egypt. Analyse the mathematical principles behind the structures, including proportions, symmetry, and stability.

15. Explore African proverbs and logic puzzles that require critical thinking and problem-solving skills. Analyse the underlying mathematical and logical concepts within the puzzles.

16. Learn about the traditional Ashanti Kente cloth weaving techniques and create your own miniature Kente cloth using paper strips. Explore the mathematical patterns created by the warp and weft threads.

17. Create a board game based on the Adrinka symbols and their meanings. Develop rules, strategies, and scoring systems that incorporate mathematical concepts such as probability and decision-making.

18. Take a virtual or in-person trip to explore African architecture and identify examples of symmetry, rotational symmetry, and reflection symmetry. Document your findings and discuss the mathematical significance of symmetry in architecture.

19. Create artwork inspired by African geometric designs using compasses, rulers, and protractors. Explore geometric constructions, such as regular polygons and geometric transformations, to create visually appealing compositions.

20. Investigate African tessellation patterns found in art and architecture. Create your own tessellations using basic shapes and explore the mathematical concepts of congruence, transformations, and repetition.

West Africa

1. Adinkra symbols from the Akan culture can be used to teach symmetry and patterns.

2. The Sankofa symbol can be used to explore polar equations and geometric shapes.

3. The Aroko number system can be used to introduce different number bases and place value.

4. The Nsimbu number system can be studied to understand alternative counting systems.

5. Gelede patterns incorporate geometric shapes and symmetry, suitable for teaching geometry.

6. Nsibidi symbols can be used to study coding and encryption techniques.

7. Traditional weaving patterns can be explored to teach symmetry, tessellations, and rotational symmetry.

8. Lattice designs from West African cultures can be used to teach transformations and patterns.
9. The Dagaaba compass can be studied to understand directions and angles.
10. Yoruba divination systems like Ifá can be used to teach probability and statistics.
11. Drumming patterns can be analysed to study rhythm, patterns, and sequences.
12. Talking drums can be used to explore sound waves and frequency.
13. The Hausa calendar can be used to teach time, calendars, and number patterns.
14. Stick calendars used by the Fulani people can be studied to understand counting, time, and patterns.
15. Ndebele art can be explored to teach geometry, symmetry, and patterns.
16. Akan goldweights can be used to teach measurement, weight, and ratios.
17. Dogon astronomy can be studied to explore celestial patterns and timekeeping.
18. Igbo Mbari art can be used to teach geometric shapes and patterns.
19. Ashanti Kente cloth patterns can be explored to teach geometry, symmetry, and patterns.
20. Mossi ironworking techniques can be used to teach measurement, angles, and shapes.
21. Senufo masks can be studied to explore symmetry, patterns, and cultural representations.
22. Wolof braiding techniques can be used to teach geometry, patterns, and symmetry.
23. Fulani counting systems can be explored to introduce different number bases and counting methods.
24. Mande sand divination can be used to teach statistics, probability, and decision-making.
25. Kuba textiles can be studied to explore geometry, patterns, and symmetry.
26. Fon geometric designs can be used to teach geometry, symmetry, and patterns.
27. Mossi loom weaving techniques can be explored to teach measurement, patterns, and symmetry.
28. The Efik Ekpe secret society can be studied to explore codes, symbols, and secret messages.
29. Tuareg nomadic geometry can be used to teach geometric shapes, patterns, and cultural practices.
30. Yoruba gele tying techniques can be explored to teach measurement, patterns, and symmetry.
31. Gurunsi earth architecture can be studied to explore geometric shapes, patterns, and cultural practices.
32. The Tukuler numeration system can be used to teach different number bases and counting methods.

33. The Igbo Igba Nkwu ceremony can be studied to explore statistics, probability, and cultural practices.
34. Guro masks can be used to teach symmetry, patterns, and cultural representations.
35. Bamana mud cloth patterns can be explored to teach geometry, patterns, and symmetry.
36. Mossi earth shrines can be studied to explore geometry, shapes, and cultural practices.
37. Temne geometric body painting techniques can be used to teach geometry, symmetry, and patterns.
38. The Dinka numeration system can be explored to teach different number bases and counting methods.
39. Yoruba Gelede masquerades can be studied to explore symmetry, patterns, and cultural practices.

North and central Africa
1. Berber weaving patterns can be explored to teach geometry, symmetry, and patterns.
2. Tuareg nomadic geometry can be used to teach geometric shapes, patterns, and cultural practices.
3. The Almoravid numeration system can be studied to understand different number bases and counting methods.
4. Arabic calligraphy can be explored to teach symmetry, patterns, and cultural representations.
5. Sudanese basket weaving techniques can be used to teach geometry, patterns, and symmetry.
6. Saharan navigation techniques can be studied to explore angles, directions, and celestial patterns.
7. Tifinagh symbols can be used to study coding and encryption techniques.
8. The Amasigh calendar can be used to teach time, calendars, and number patterns.
9. Saharan astronomy can be studied to explore celestial patterns and timekeeping.
10. Coptic geometric art can be used to teach geometry, symmetry, and patterns.
11. Sudanese architectural designs can be explored to teach geometric shapes, patterns, and symmetry.
12. The Nubian number system can be studied to understand different number bases and counting methods.
13. Swahili door carvings can be used to teach symmetry, patterns, and cultural representations.
14. Aksumite architecture can be explored to teach geometry, shapes, and patterns.

15. Ethiopian crosses can be studied to explore symmetry, patterns, and cultural representations.
16. Ancient Egyptian mathematics, such as the Rhind Mathematical Papyrus, can be used to teach various topics, including geometry, fractions, and arithmetic.
17. Sudanese pyramids can be explored to teach geometry, shapes, and measurement.
18. Bantu divination systems can be studied to explore patterns, probabilities, and cultural practices.
19. Dogon symbolic systems can be used to teach coding, patterns, and cultural representations.
20. Khoisan click languages can be explored to study patterns, sequences, and linguistic structures.
21. Nok terracotta sculptures can be studied to explore geometric shapes, patterns, and cultural representations.
22. Swahili architectural designs can be used to teach geometry, shapes, and patterns.
23. San rock art can be explored to study symmetry, patterns, and cultural representations.
24. Ethiopian cross knotwork can be used to teach geometry, symmetry, and patterns.
25. Nubian pottery designs can be studied to explore geometric shapes, patterns, and symmetry.
26. Moroccan sellige tiles can be used to teach symmetry, patterns, and geometric transformations.
27. Berber nomadic counting systems can be explored to introduce different number bases and counting methods.
28. Sudanese mud brick architecture can be studied to explore geometry, shapes, and cultural practices.
29. Tuareg metalwork techniques can be used to teach measurement, shapes, and geometric transformations.
30. Ndebele house paintings can be explored to teach symmetry, patterns, and cultural representations.
31. Sahelian agriculture techniques can be studied to explore geometric patterns, proportions, and measurement.
32. Mossi pottery designs can be used to teach geometry, patterns, and symmetry.
33. Swahili wood carvings can be explored to teach symmetry, patterns, and cultural representations.
34. Hausa geometric patterns can be used to teach geometry, symmetry, and patterns.
35. Egyptian hieroglyphics can be studied to explore coding, patterns, and cultural representations.
36. Coptic weaving techniques can be used to teach measurement, patterns, and symmetry.

37. Benin bronze plaques can be explored to teach geometry, patterns, and cultural representations.
38. Sudanese geodesic domes can be studied to explore geometric shapes, patterns, and cultural practices.
39. Nubian tomb architecture can be used to teach geometry, shapes, and patterns.
40. Ethiopian illuminated manuscripts can be explored to study symmetry, patterns, and cultural representations.

East Africa
1. Swahili geometric patterns can be explored to teach geometry, symmetry, and patterns.
2. Maasai beadwork can be used to teach patterns, sequences, and cultural representations.
3. Akamba woodcarving techniques can be studied to explore symmetry, patterns, and cultural representations.
4. Somali navigation techniques can be used to teach angles, directions, and celestial patterns.
5. Lamu architecture can be explored to teach geometry, shapes, and patterns.
6. The Kikuyu Mundu number system can be studied to understand different number bases and counting methods.
7. Kamba sand drawings can be used to teach geometry, shapes, and patterns.
8. Samburu counting systems can be explored to introduce different number bases and counting methods.
9. Kikuyu crop rotation techniques can be studied to explore patterns, sequences, and agricultural practices.
10. The Giriama calendar can be used to teach time, calendars, and number patterns.
11. Turkana nomadic geometry can be explored to teach geometric shapes, patterns, and cultural practices.
12. Rendille geometric designs can be studied to explore geometry, symmetry, and patterns.
13. Chagga terraced farming techniques can be used to teach measurement, geometry, and agricultural practices.
14. Meru basket weaving techniques can be studied to explore geometry, patterns, and symmetry.
15. Swahili coral stone architecture can be used to teach geometry, shapes, and patterns.
16. Kipsigis rainmaking techniques can be explored to introduce statistics, probability, and cultural practices.
17. Digo traditional fishing techniques can be studied to explore measurement, proportions, and cultural practices.

18. Luo fishing nets can be used to teach measurement, patterns, and cultural representations.
19. Rusinga Island megaliths can be explored to study measurement, shapes, and cultural practices.
20. Taita basket weaving techniques can be used to teach geometry, patterns, and symmetry.
21. The Borana calendar can be studied to explore time, calendars, and number patterns.
22. Makonde wood carvings can be explored to study symmetry, patterns, and cultural representations.
23. Turkana astronomy can be used to explore celestial patterns, timekeeping, and cultural practices.
24. Luo musical instruments can be studied to explore patterns, sequences, and cultural representations.
25. Kamba architectural designs can be used to teach geometry, shapes, and patterns.
26. Chagga banana fibres can be explored to teach measurement, proportions, and cultural practices.
27. Kikuyu drumming patterns can be used to teach patterns, sequences, and cultural representations.
28. Kuria divination systems can be studied to introduce statistics, probability, and cultural practices.
29. Rendille camel tracking techniques can be used to teach measurement, geometry, and cultural practices.
30. Maasai shuka patterns can be explored to study symmetry, patterns, and cultural representations.
31. Kamba reed mat weaving techniques can be used to teach measurement, patterns, and symmetry.
32. Giriama crop harvesting techniques can be studied to explore measurement, proportions, and agricultural practices.
33. Maasai cow herding practices can be used to teach measurement, patterns, and cultural practices.
34. Taita rock art can be explored to study symmetry, patterns, and cultural representations.
35. Swahili dhows can be used to teach measurement, geometry, and cultural representations.
36. Kamba basket weaving techniques can be studied to explore geometry, patterns, and symmetry.
37. Samburu beaded jewellery can be used to teach patterns, sequences, and cultural representations.
38. Kikuyu hut building techniques can be explored to teach geometry, shapes, and cultural practices.

Southern Africa

1. Ndebele wall art can be explored to teach symmetry, patterns, and cultural representations.
2. San rock paintings can be used to study symmetry, patterns, and cultural representations.
3. Sulu beadwork can be studied to explore patterns, sequences, and cultural representations.
4. Venda pottery designs can be used to teach geometry, patterns, and cultural representations.
5. Tsonga basket weaving techniques can be explored to teach measurement, patterns, and symmetry.
6. Xhosa beaded jewellery can be used to teach patterns, sequences, and cultural representations.
7. Shona stone sculptures can be studied to explore geometry, shapes, and cultural representations.
8. Tswana geometric patterns can be used to teach geometry, symmetry, and patterns.
9. San hunting techniques can be explored to introduce statistics, probabilities, and cultural practices.
10. Nguni cow herding practices can be used to teach measurement, patterns, and cultural practices.
11. Sotho wall murals can be studied to explore symmetry, patterns, and cultural representations.
12. Swazi reed dance can be used to teach statistics, probabilities, and cultural practices.
13. Venda woodcarving techniques can be explored to teach geometry, patterns, and cultural representations.
14. Nama quilt patterns can be used to teach symmetry, patterns, and cultural representations.
15. Sulu shield designs can be used to teach geometry, symmetry, and patterns.
16. San storytelling techniques can be explored to introduce patterns, sequences, and cultural practices.
17. Shangaan drumming patterns can be used to teach patterns, sequences, and cultural representations.
18. Ndebele neck rings can be studied to explore measurement, proportions, and cultural practices.
19. Tsonga reed dance can be used to teach statistics, probabilities, and cultural practices.
20. Venda copper sculptures can be explored to teach geometry, shapes, and cultural representations.

21. Sotho Basotho blankets can be used to teach symmetry, patterns, and cultural representations.
22. San tracking techniques can be studied to introduce measurement, proportions, and cultural practices.
23. Tswana basket weaving techniques can be used to teach geometry, patterns, and symmetry.
24. Sulu stick fighting techniques can be explored to teach measurement, angles, and cultural practices.
25. Shona hut construction techniques can be used to teach geometry, shapes, and cultural practices.
26. Nama rock engravings can be studied to explore symmetry, patterns, and cultural representations.
27. Xhosa beaded bracelets can be used to teach patterns, sequences, and cultural representations.
28. Venda tribal mask designs can be explored to teach geometry, patterns, and cultural representations.
29. Sotho Lesiba music can be used to teach patterns, sequences, and cultural representations.
30. Tsonga drumming techniques can be studied to explore patterns, sequences, and cultural representations.
31. San Bushman fire-making techniques can be used to teach measurement, proportions, and cultural practices.
32. Sulu reed mat weaving techniques can be explored to teach geometry, patterns, and symmetry.
33. Ndebele colours symbolism can be used to study statistics, probabilities, and cultural representations.
34. Shona copper wire sculptures can be explored to teach geometry, shapes, and cultural representations.
35. Venda musical instruments can be used to teach patterns, sequences, and cultural representations.
36. Tsonga marula fruit harvesting techniques can be studied to explore measurement, proportions, and cultural practices.
37. Sotho Basotho horse riding practices can be used to teach measurement, patterns, and cultural practices.
38. San basket weaving techniques can be explored to teach geometry, patterns, and symmetry.
39. Nama Astronomy: Nama astronomy can be used to study celestial patterns, timekeeping, and cultural practices.

Caribbean

1. Caribbean carnival costume designs can be explored to teach geometry, patterns, and cultural representations.
2. Rastafarian dreadlock techniques can be used to teach measurement, proportions, and cultural practices.
3. Calypso music rhythms can be studied to introduce patterns, sequences, and cultural representations.
4. Caribbean steelpan tuning techniques can be used to teach measurement, proportions, and cultural practices.
5. Jamaican dub poetry can be explored to teach patterns, sequences, and cultural representations.
6. Trinidadian limbo dance can be used to study angles, measurement, and cultural practices.
7. Jamaican riddim patterns can be used to teach patterns, sequences, and cultural representations.
8. Barbadian Crop Over festival can be explored to introduce statistics, probabilities, and cultural practices.
9. Dominican merengue dance can be studied to explore patterns, sequences, and cultural representations.
10. Cuban dominoes game strategies can be used to teach problem-solving, logical reasoning, and cultural practices.
11. Trinidadian carnival mask designs can be explored to teach geometry, patterns, and cultural representations.
12. Jamaican Maroon drumming techniques can be used to teach patterns, sequences, and cultural representations.
13. Trinidadian chutney music rhythms can be used to teach patterns, sequences, and cultural representations.
14. Jamaican Patois counting system can be studied to understand different number systems and cultural practices.
15. Cuban salsa dance steps can be used to teach patterns, sequences, and cultural representations.
16. Jamaican Ital cooking ratios can be used to teach proportions, measurements, and cultural practices.
17. Trinidadian steelband arrangements can be studied to explore patterns, sequences, and cultural representations.
18. Dominican bachata dance moves can be used to teach patterns, sequences, and cultural representations.

19. Jamaican dancehall steps can be used to teach patterns, sequences, and cultural representations.
20. Bahamian straw weaving techniques can be studied to teach geometry, patterns, and cultural representations.
21. Guyanese pepperpot recipe ratios can be used to teach proportions, measurements, and cultural practices.
22. Trinidadian soca music rhythms can be explored to teach patterns, sequences, and cultural representations.
23. Cuban mambo dance patterns can be used to teach patterns, sequences, and cultural representations.
24. Barbadian crop rotation techniques can be studied to explore measurement, proportions, and agricultural practices.
25. Jamaican Nyabinghi drumming techniques can be used to teach patterns, sequences, and cultural representations.
26. Trinidadian doubles recipe ratios can be explored to teach proportions, measurements, and cultural practices.
27. Puerto Rican bomba dance movements can be used to teach patterns, sequences, and cultural representations.
28. Cuban cachibache craft making techniques can be studied to explore geometry, patterns, and cultural representations.
29. Jamaican Maroon herbal medicine practices can be used to teach measurement, proportions, and cultural practices.
30. Bahamian Junkanoo costumes can be explored to teach symmetry, patterns, and cultural representations.
31. Trinidadian parang music harmonies can be used to teach patterns, sequences, and cultural representations.
32. Jamaican Dub Poetry: Jamaican dub poetry can be explored to teach patterns, sequences, and cultural representations.
33. Jamaican Nyamings Counting Rhymes: Jamaican Nyamings counting rhymes can be used to teach patterns, sequences, and cultural representations.

Asia

1. Indian Rangoli designs can be explored to teach geometry, symmetry, and patterns.
2. Pakistani truck art patterns can be used to teach geometry, shapes, and cultural representations.
3. Bangladeshi Jamdani saree designs can be used to teach geometry, patterns, and cultural representations.

4. Nepalese mandala art can be explored to teach geometry, symmetry, and cultural representations.

5. Bhutanese Thangka paintings can be used to study geometry, proportions, and cultural representations.

6. Maldivian coral reef measurement techniques can be studied to explore geometry, proportions, and environmental conservation.

7. Indian classical dance movements can be used to teach patterns, sequences, and cultural representations.

8. Pakistani henna designs can be explored to teach geometry, symmetry, and cultural representations.

9. Sri Lankan traditional drumming patterns can be used to teach patterns, sequences, and cultural representations.

10. Bangladeshi rickshaw art can be studied to explore geometry, shapes, and cultural representations.

11. Nepalese prayer wheel calculations can be used to introduce measurement, circumference, and cultural practices.

12. Bhutanese archery techniques can be explored to teach measurement, angles, and cultural practices.

13. Maldivian Dhoni navigation techniques can be used to study geometry, navigation, and cultural practices.

14. Indian sari draping techniques can be explored to teach measurement, proportions, and cultural practices.

15. Pakistani Mughal architecture can be used to teach geometry, symmetry, and cultural representations.

16. Sri Lankan Ayurvedic medicine ratios can be studied to introduce proportions, measurements, and traditional medicine practices.

17. Bangladeshi traditional boat building techniques can be used to teach geometry, measurements, and cultural practices.

18. Nepalese prayer beads counting can be explored to teach number systems, cultural practices, and mindfulness.

19. Bhutanese traditional song rhythms can be used to teach patterns, sequences, and cultural representations.

20. Maldivian Moon Calendar Calculations: Maldivian moon calendar calculations can be studied to introduce calendars, timekeeping, and cultural practices.

21. Bangladeshi bricklaying patterns can be studied to explore geometry, patterns, and architectural practices.

22. Nepalese Thanka painting proportions can be used to teach measurements, proportions, and cultural representations.

23. Bhutanese prayer flag geometry can be explored to teach geometry, symmetry, and cultural practices.
24. Maldivian fish market price analysis can be used to introduce statistics, data analysis, and economic studies.
25. Indian classical music raga scales can be studied to explore patterns, sequences, and cultural representations.
26. Pakistani mosaic art can be used to teach geometry, shapes, and cultural representations.
27. Lankan traditional mask making techniques can be explored to teach geometry, patterns, and cultural representations.
28. Bangladeshi traditional pottery techniques can be used to teach measurements, shapes, and cultural practices.
29. Nepalese Buddhist stupa geometry can be studied to introduce geometry, symmetry, and cultural representations.
30. Bhutanese archery scorekeeping can be used to teach numbers, statistics, and cultural practices.
31. Maldivian Dhoni sail design can be explored to teach geometry, shapes, and cultural representations.
32. Indian henna measurements can be used to introduce measurement, proportions, and cultural practices.
33. Sri Lankan traditional cooking ratios can be used to teach proportions, measurements, and cultural practices.
34. Bangladeshi jute weaving patterns can be explored to teach geometry, patterns, and cultural representations.
35. Bhutanese traditional costume design can be studied to explore geometry, patterns, and cultural representations.
36. Maldivian coral stone construction techniques can be used to teach geometry, measurements, and cultural practices.
37. Indian Kolam art can be explored to teach geometry, symmetry, and cultural representations.
38. Sri Lankan Traditional Calendar Calculations: Sri Lankan traditional calendar calculations can be studied to introduce calendars, timekeeping, and cultural practices.
39. Nepalese mandala geometry can be used to teach geometry, symmetry, and cultural representations.
40. Pakistani traditional carpet weaving techniques can be used to teach measurements, patterns, and cultural practices.

Eastern Europe

1. Russian Nesting Dolls (Matryoshka): Russian nesting dolls can be used to teach proportions, ratios, and spatial reasoning.

2. Polish Folk Paper Cutting (Wycinanki): Polish folk paper cutting patterns can be explored to teach symmetry, geometry, and cultural representations.

3. Ukrainian Easter Egg Decorating (Pysanky): Ukrainian Easter egg decorating techniques can be studied to introduce symmetry, patterns, and cultural representations.

4. Romanian traditional weaving techniques can be used to teach patterns, geometry, and cultural practices.

5. Hungarian folk dance rhythms can be explored to teach patterns, sequences, and cultural representations.

6. Bulgarian embroidery designs can be used to study symmetry, patterns, and cultural representations.

7. Slovakian gingerbread mold shapes can be explored to teach geometry, shapes, and cultural representations.

8. Czech Traditional Paper Folding (Moravian Stars): Czech traditional paper folding can be used to introduce geometry, symmetry, and cultural representations.

9. Russian dacha gardening measurements can be studied to teach measurement, area, and gardening practices.

10. Polish folk music rhythmic notation can be explored to teach patterns, sequences, and cultural representations.

11. Ukrainian traditional costumes can be used to study geometry, shapes, and cultural representations.

12. Romanian traditional folk dances can be explored to teach patterns, sequences, and cultural representations.

13. Hungarian paprika harvest measurements can be used to introduce measurement, estimation, and agricultural practices.

14. Bulgarian traditional pottery patterns can be studied to explore geometry, patterns, and cultural representations.

15. Slovakian traditional folk songs can be used to teach patterns, sequences, and cultural representations.

16. Czech traditional folk tales can be explored to teach problem-solving, logical reasoning, and cultural mathematics.

17. Russian iconography symmetry can be used to introduce symmetry, geometry, and cultural representations.

18. Polish traditional cheese making techniques can be studied to teach measurement, proportions, and cultural practices.

19. Ukrainian traditional village architecture can be explored to teach geometry, shapes, and cultural representations.
20. Romanian traditional folklore number games can be used to teach problem-solving, logical reasoning, and cultural mathematics.
21. Hungarian folk costume embroidery can be studied to explore geometry, patterns, and cultural representations.
22. Bulgarian traditional honey production techniques can be used to teach measurement, proportions, and cultural practices.
23. Slovakian traditional folk instrument tuning can be explored to introduce ratios, frequencies, and cultural practices.
24. Czech astronomical clock mechanics can be used to teach time, gears, and cultural representations.
25. Russian samovar heat transfer can be studied to introduce thermodynamics, heat conduction, and cultural practices.
26. Polish traditional folklore calendar customs can be explored to teach calendars, timekeeping, and cultural practices.
27. Ukrainian traditional wedding rituals can be used to study numbers, counting, and cultural representations.
28. Romanian traditional wood carving techniques can be explored to teach geometry, shapes, and cultural representations.
29. Hungarian traditional folk games can be used to teach problem-solving, logical reasoning, and cultural mathematics.
30. Bulgarian traditional rug weaving techniques can be studied to introduce patterns, geometry, and cultural representations.
31. Slovakian Easter whip measurement can be explored to teach measurement, length, and cultural practices.
32. Czech traditional folk music instruments can be used to study sound waves, frequencies, and cultural representations.
33. Russian tea brewing ratios can be studied to introduce proportions, measurement, and cultural practices.
34. Polish traditional Easter basket arrangement can be explored to teach geometry, symmetry, and cultural representations.
35. Ukrainian traditional embroidery stitching techniques can be used to teach patterns, geometry, and cultural representations.
36. Romanian traditional ceramics can be studied to explore geometry, shapes, and cultural representations.
37. Hungarian traditional folk poetry can be used to teach problem-solving, logical reasoning, and cultural mathematics.

38. Bulgarian traditional bread baking techniques can be explored to teach measurement, proportions, and cultural practices.
39. Slovakian traditional architecture can be used to study geometry, shapes, and cultural representations.
40. Czech traditional puppetry geometry can be studied to introduce geometry, shapes, and cultural representations.
41. Russian constructivist art proportions can be explored to teach geometry, proportions, and cultural representations.
42. Polish traditional folk medicine practices can be used to introduce measurement, ratios, and cultural practices.
43. Ukrainian traditional Pysanky geometry can be studied to introduce symmetry, patterns, and cultural representations.
44. Romanian traditional carpet weaving techniques can be explored to teach geometry, patterns, and cultural representations.
45. Hungarian traditional folk motifs can be used to study geometry, symmetry, and cultural representations.
46. Bulgarian traditional wine making techniques can be explored to teach measurement, proportions, and cultural practices.
47. Slovakian traditional folklore calendar systems can be used to introduce calendars, timekeeping, and cultural practices.
48. Czech traditional wood joinery techniques can be studied to explore geometry, shapes, and cultural representations.
49. Russian Orthodox cross symmetry can be used to teach symmetry, geometry, and cultural representations.
50. Polish traditional folk costume measurements can be explored to teach measurement, proportions, and cultural practices.
51. Ukrainian traditional Cossack dancing can be used to teach patterns, sequences, and cultural representations.
52. Bulgarian traditional wedding customs can be explored to teach numbers, counting, and cultural representations.
53. traditional folklore clothing can be used to study geometry, shapes, and cultural representations.
54. Czech traditional glass blowing techniques can be explored to teach measurement, geometry, and cultural practices.
55. Russian traditional lacquer miniature painting can be used to introduce symmetry, patterns, and cultural representations.
56. Polish traditional folklore knot patterns can be studied to explore geometry, patterns, and cultural representations.

57. Ukrainian traditional Hutsul wood carving techniques can be used to teach geometry, shapes, and cultural representations.
58. Romanian traditional musical instruments can be explored to study sound waves, frequencies, and cultural representations.

Western Europe

1. Greek Geometry: The study of Greek geometry, including the Pythagorean theorem ($a^2 + b^2 = c^2$), can be used to teach geometric principles, proofs, and the relationship between the sides of a right-angled triangle.
2. Roman numerals can be explored to teach number systems, place value, and historical number representations.
3. The use of the golden ratio (1.618...) in Italian Renaissance art can be studied to introduce proportion, geometry, and aesthetic principles.
4. Measurement conversions used in French boulangeries (e.g., converting grams to ounces or milliliters to cups) can be used to teach measurement, conversion, and practical applications of mathematics.
5. The British currency system, including pounds (£) and pence (p), can be explored to teach decimals, money calculations, and financial literacy.
6. German clockmaking traditions, such as cuckoo clocks, can be used to teach time, intervals, and mechanisms.
7. Spanish flamenco dance rhythms can be studied to explore patterns, sequences, and cultural representations.
8. Portuguese tile patterns can be used to study geometry, symmetry, and cultural representations.
9. Irish Celtic knots can be explored to teach geometry, patterns, and cultural representations.
10. Dutch windmill sail angle calculations can be used to introduce trigonometry, angles, and mechanical principles.
11. SSwiss watchmaking precision can be studied to explore measurement, timekeeping, and precision engineering.
12. Belgian chocolate box perimeter and area calculations can be used to teach geometry, measurement, and real-life applications.
13. Norwegian Viking navigation methods, such as using the sunstone for celestial navigation, can be explored to teach angles, navigation, and historical mathematics.
14. Danish Lego building can be used to develop spatial reasoning, geometry, and problem-solving skills.

15. Scottish tartan pattern symmetry can be studied to explore geometry, symmetry, and cultural representations.
16. Finnish sauna temperature conversions (Celsius to Fahrenheit) can be used to teach measurement, temperature scales, and conversion.
17. The Viennese waltz can be explored to teach rhythm, patterns, and cultural representations.
18. Swedish furniture assembly diagrams can be used to teach spatial reasoning, geometry, and problem-solving skills.
19. Icelandic sagas can be studied to create word problems, develop critical thinking, and apply mathematical concepts in historical contexts.
20. Swiss Alps elevation and gradient calculations can be used to teach measurement, slopes, and geographical mathematics.
21. Welsh Celtic cross symmetry can be explored to teach geometry, symmetry, and cultural representations.
22. Italian pizza dough ratios (flour to water to yeast) can be studied to introduce ratios, proportions, and culinary mathematics.
23. French vineyard harvest yield calculations can be used to teach measurement, estimation, and agricultural mathematics.
24. Spanish bullfighting geometry, such as the angles formed by the matador and the bull, can be explored to teach angles, shapes, and cultural representations.
25. German Oktoberfest beer stein volume calculations can be used to teach measurement, capacity, and real-life applications.
26. The British Royal Guard changing of the guard ceremony can be studied to teach time, intervals, and precision.
27. Dutch tulip field symmetry can be explored to teach geometry, symmetry, and cultural representations.
28. Irish step dance counting and rhythm can be used to teach patterns, sequences, and cultural representations.
29. : Swiss chocolate bar fraction divisions can be studied to introduce fractions, division, and real-life applications.
30. Belgian waffle grid patterns can be used to explore geometry, symmetry, and cultural representations.
31. Italian opera music time signatures can be studied to introduce rhythm, fractions, and cultural representations.
32. French fashion design and proportional scaling can be used to teach ratios, scaling, and real-life applications.
33. Spanish mosaic tiles and tessellations can be studied to introduce geometry, patterns, and cultural representations.

34. Portuguese asulejo tile surface area calculations can be used to teach geometry, measurement, and cultural representations.
35. Irish Gaelic language numerical systems can be explored to teach number systems, counting, and cultural representations.
36. Dutch canal lock water level calculations can be used to teach measurement, volume, and engineering mathematics.
37. Scottish bagpipe sound waves can be studied to explore sound waves, frequencies, and cultural representations.
38. Norwegian Sami traditional reindeer herding can be used to introduce measurement, estimation, and cultural practices.
39. Swiss cheese hole area calculations can be studied to teach geometry, measurement, and real-life applications.
40. Italian fresco painting perspective can be explored to teach geometry, depth perception, and cultural representations.
41. French wine vineyard grape harvest estimation can be used to teach measurement, estimation, and agricultural mathematics.
42. German fairy tale castles and architectural scale models can be studied to introduce ratios, scaling, and cultural representations.
43. Spanish flamenco guitar chord progressions can be used to teach patterns, sequences, and cultural representations.
44. Portuguese fado music melody intervals can be explored to introduce intervals, scales, and cultural representations.
45. Irish Celtic harp string length ratios can be used to study proportions, harmonic series, and cultural representations.
46. Dutch Delft Blue pottery symmetry can be explored to teach geometry, symmetry, and cultural representations.
47. Scottish Highland Games measurement conversions (e.g., tossing the caber) can be used to teach measurement, conversion, and practical applications of mathematics.
48. Finnish Northern Lights geometry can be studied to explore shapes, patterns, and cultural representations.
49. Swiss Alpine horn sound waves can be used to teach sound waves, frequencies, and cultural representations.
50. Welsh Eisteddfod poetry meter and rhythm can be explored to introduce poetic meters, rhythm, and cultural representations.
51. Italian gelato recipe scaling can be used to teach ratios, scaling, and culinary mathematics.
52. The French Revolution timeline and data analysis can be studied to introduce timelines, data representation, and historical mathematics.

53. German Christmas markets profit calculations can be used to teach financial mathematics, profit margins, and real-life applications.
54. Spanish bull run speed calculations can be explored to teach speed, distance, and real-life applications.
55. Portuguese fado song length proportions can be used to study ratios, proportions, and cultural representations.
56. Irish pub session musicians and time signatures can be explored to introduce rhythm, fractions, and cultural representations.
57. Dutch cycling distance and time rate calculations can be used to teach speed, time, and real-life applications.
58. Scottish tartan colours pattern combinations can be studied to explore permutations, combinations, and cultural representations.
59. Finnish sauna temperature time graphs can be used to introduce graphs, data representation, and cultural practices.

East, West, and North Asia

The Chinese abacus can be used to teach place value, addition, subtraction, and multiplication.

Japanese origami can be explored to teach geometry, spatial reasoning, and symmetry.

1. Vedic mathematics techniques, such as multiplication by vertically and crosswise, can be used to teach mental calculation and speed mathematics.
2. Korean Geobukseon turtle ship symmetry can be studied to explore geometry, symmetry, and historical representations.
3. Mongolian horseback archery trajectory calculations can be used to teach angles, projectile motion, and real-life applications.
4. Thai Buddhist temple architecture proportions can be explored to teach ratios, scaling, and cultural representations.
5. Vietnamese Dong Son drum patterns can be studied to explore patterns, sequences, and cultural representations.
6. Indonesian batik textile symmetry can be used to teach geometry, symmetry, and cultural representations.
7. Japanese tea ceremony time measurement techniques can be explored to teach time, intervals, and precision.
8. Chinese Tangram puzzles can be used to teach geometry, area, perimeter, and problem-solving skills.
9. Indian Rangoli patterns can be studied to explore geometry, symmetry, and cultural representations.

10. Korean ssireum wrestling force and balance calculations can be used to teach weight, force, and equilibrium.
11. Japanese shakuhachi flute sound waves can be explored to teach sound waves, frequencies, and cultural representations.
12. Chinese Feng Shui principles can be studied to explore geometry, spatial arrangement, and cultural practices.
13. Indian Kolam patterns can be used to teach geometry, symmetry, and cultural representations.
14. Korean Janggi strategic thinking can be explored to teach problem-solving, critical thinking, and logical reasoning.
15. Japanese Haiku poetry syllable counting can be used to teach counting, patterns, and cultural representations.
16. Chinese dragon boat racing timekeeping can be studied to introduce time, intervals, and precision.
17. Indian sitar music scale intervals can be explored to introduce intervals, scales, and cultural representations.
18. Mongolian ger yurt construction proportions can be used to teach geometry, scaling, and cultural representations.
19. Thai Muay Thai boxing weight category calculations can be studied to introduce measurement, weight conversions, and sports mathematics.
20. Vietnamese water puppetry puppet motion trajectories can be explored to teach angles, trajectories, and cultural representations.
21. Indonesian Gamelan ensemble rhythm and patterns can be used to teach patterns, sequences, and cultural representations.
22. Japanese Noh theatre masks can be studied to explore geometry, symmetry, and cultural representations.
23. Chinese martial arts stances can be used to teach trigonometric concepts, angles, and real-life applications.
24. Indian yoga poses and body proportions can be explored to teach geometry, proportions, and cultural practices.
25. Korean fan dance counting and rhythm can be used to teach patterns, sequences, and cultural representations.
26. Japanese bento box fraction calculations can be studied to introduce fractions, division, and cultural representations.
27. Chinese zodiac animal year cycles can be explored to teach number patterns, cycles, and cultural representations.
28. Indian classical dance hand gestures can be used to teach geometry, symmetry, and cultural representations.

29. Thai Buddhist temple bell frequencies can be studied to explore sound waves, harmonics, and cultural representations.
30. Vietnamese Ao Dai clothing design and measurement can be used to teach geometry, measurement, and cultural representations.
31. Indonesian Wayang shadow puppetry shadow length proportions can be explored to teach ratios, proportions, and cultural representations.
32. Japanese Ikebana flower arrangement symmetry can be studied to introduce geometry, symmetry, and cultural representations.
33. Chinese lantern festival surface area calculations can be used to teach geometry, measurement, and cultural representations.
34. Indian classical music ragas can be explored to introduce intervals, scales, and cultural representations.
35. Korean fan dance area and perimeter calculations can be used to teach geometry, area, perimeter, and problem-solving skills.
36. Japanese Sen garden rock arrangements can be studied to explore patterns, sequences, and cultural representations.
37. Chinese kite geometry and symmetry can be used to teach geometry, symmetry, and cultural representations.
38. Indian henna design reflection symmetry can be explored to teach geometry, symmetry, and cultural representations.
39. Mongolian throat singing harmonics can be studied to explore sound waves, harmonics, and cultural representations.
40. Thai fruit carving geometric patterns can be used to teach geometry, symmetry, and cultural representations.
41. Vietnamese Ao Dai dress patterns and measurement can be explored to teach geometry, measurement, and cultural representations.
42. Indonesian batik wax resist techniques can be used to introduce fractions, ratios, and cultural practices.
43. Japanese Maneki-neko lucky cat patterns can be studied to explore geometry, patterns, and cultural representations.
44. Chinese lion dance choreography counting can be used to teach patterns, counting, and cultural representations.
45. Indian Holi festival colours ratios can be explored to introduce ratios, proportions, and cultural representations.
46. Korean K-pop choreography time signatures can be studied to teach rhythm, fractions, and cultural representations.
47. Japanese samurai sword cutting angles can be used to teach angles, measurement, and historical mathematics.

48. Chinese calligraphy brush stroke sequences can be explored to teach patterns, sequences, and cultural representations.

49. Thai Songkran water festival volume and capacity calculations can be explored to teach measurement, volume, and cultural representations.

50. Vietnamese water puppetry puppet manipulation angles can be studied to introduce angles, geometry, and cultural representations.

51. Indonesian shadow puppetry Wayang Kulit area and perimeter calculations can be used to teach geometry, area, perimeter, and problem-solving skills.

52. Japanese tea ceremony tea ratios and proportions can be explored to introduce ratios, proportions, and cultural practices.

53. Chinese lantern festival symmetry and reflection can be studied to explore geometry, symmetry, and cultural representations.

54. Indian classical dance mudras can be used to teach geometry, shapes, and cultural representations.

55. Korean royal cuisine ingredient measurement conversions can be explored to teach measurement, conversions, and culinary mathematics.

56. Japanese Haiku poetry seasonal word associations can be studied to introduce word associations, patterns, and cultural representations.

North and South America

1. The Maya numerical system, which uses a base-20 system, can be used to teach place value, addition, and multiplication.

2. Inca quipu knots can be explored to teach data representation, patterns, and cultural representations.

3. Native American medicine wheel geometry can be studied to explore circles, angles, and cultural representations.

4. The Aztec calendar stone symmetry can be used to teach symmetry, patterns, and cultural representations.

5. Native American petroglyph patterns can be explored to teach geometry, patterns, and cultural representations.

6. Mayan astronomical observations can be studied to introduce angles, measurement, and historical mathematics.

7. Native American tipi conical geometry can be used to teach geometry, measurement, and cultural representations.

8. Aztec floating gardens area calculations can be explored to teach geometry, measurement, and historical mathematics.

9. Native American dreamcatcher patterns can be studied to explore geometry, symmetry, and cultural representations.
10. Inuit igloo volume and surface area calculations can be used to teach geometry, measurement, and cultural representations.
11. Native American beadwork patterns can be explored to teach patterns, sequences, and cultural representations.
12. Aztec chocolate trade barter system calculations can be studied to introduce currency, conversions, and historical mathematics.
13. Native American storytelling numeric patterns can be used to teach number patterns, sequences, and cultural representations.
14. Mayan hieroglyphic numerals can be explored to introduce numerals, place value, and historical mathematics.
15. Inca terrace farming slope calculations can be studied to teach angles, measurement, and agricultural mathematics.
16. Native American counting sticks and bundles can be used to teach place value, counting, and cultural representations.
17. Native American Navajo Rug Patterns: Native American Navajo rug patterns can be studied to introduce geometry, patterns, and cultural representations.
18. Inuit Snow Goggles Refraction and Light: Inuit snow goggles refraction and light calculations can be used to teach optics, angles, and cultural representations.
19. Native American Medicine Men Healing Time Intervals: Native American medicine men healing time intervals can be explored to teach time, intervals, and cultural practices.
20. Aztec Temple Architecture Scaling: Aztec temple architecture scaling can be studied to introduce ratios, scaling, and historical representations.
21. Native American Hopi Kachina Doll Symmetry: Native American Hopi Kachina doll symmetry can be used to teach symmetry, patterns, and cultural representations.
22. Inca Road Networks Distance and Time Calculations: Inca road networks distance and time calculations can be explored to teach distance, time, and historical mathematics.
23. Native American Pottery Volume and Capacity: Native American pottery volume and capacity calculations can be used to teach geometry, measurement, and cultural representations.
24. Aztec Featherwork Geometric Designs: Aztec featherwork geometric designs can be studied to explore geometry, patterns, and cultural representations.
25. Native American Sun Dance Counting and Rhythm: Native American sun dance counting and rhythm can be used to teach patterns, counting, and cultural representations.
26. Inuit Kayak Design and Buoyancy: Inuit kayak design and buoyancy calculations can be explored to teach geometry, physics, and cultural representations.

27. **Native American Cherokee Beadwork Symmetry:** Native American Cherokee beadwork symmetry can be studied to introduce symmetry, patterns, and cultural representations.

28. **Aztec Market Economy Money Calculations:** Aztec market economy money calculations can be used to teach currency, conversions, and historical mathematics.

29. **Native American Wampum Belts Binary Systems:** Native American Wampum belts binary systems can be explored to introduce binary numbers, counting, and cultural representations.

30. **Inca Stone Masonry Angles and Proportions:** Inca stone masonry angles and proportions can be studied to teach geometry, angles, and historical mathematics.

31. **Native American Pueblo Architecture Ratios:** Native American Pueblo architecture ratios can be used to teach ratios, proportions, and cultural representations.

32. **Aztec Gold Alloy Mixtures:** Aztec gold alloy mixtures calculations can be explored to teach ratios, fractions, and historical mathematics.

33. **Native American Lakota Star Quilt Symmetry:** Native American Lakota star quilt symmetry can be studied to introduce symmetry, patterns, and cultural representations.

34. **Inuit Northern Lights Patterns:** Inuit Northern Lights patterns can be used to teach patterns, sequences, and cultural representations.

35. **Native American Plains Indians Bison Hunting Statistics:** Native American Plains Indians bison hunting statistics can be explored to teach data analysis, statistics, and cultural practices.

36. **Aztec Geometry of Ball Courts:** Aztec geometry of ball courts can be studied to introduce angles, measurement, and historical mathematics.

37. **Native American Iroquois Longhouse Measurement:** Native American Iroquois longhouse measurement can be used to teach geometry, measurement, and cultural representations.

38. **Inca Textile Weaving Patterns and Symmetry:** Inca textile weaving patterns can be explored to teach patterns, symmetry, and cultural representations.

39. **Native American Powwow Drum Rhythms:** Native American powwow drum rhythms can be studied to introduce rhythm, fractions, and cultural representations.

40. **Aztec Codex Glyphs and Numerals:** Aztec codex glyphs and numerals can be used to teach numerals, symbols, and historical mathematics.

41. **Native American Tlingit Totem Pole Heights:** Native American Tlingit totem pole heights can be explored to teach measurement, height, and cultural representations.

42. **Inuit Inukshuk Construction and Proportions:** Inuit inukshuk construction and proportions can be studied to introduce geometry, proportions, and cultural representations.

43. **Native American Choctaw Stickball Game Time Intervals:** Native American Choctaw stickball game time intervals can be used to teach time, intervals, and cultural practices.

44. **Aztec Agriculture Calendar Planting Seasons:** Aztec agriculture calendar planting seasons can be explored to introduce calendars, seasons, and historical mathematics.

45. Native American Mohawk Basket Weaving Patterns: Native American Mohawk basket weaving patterns can be studied to explore geometry, patterns, and cultural representations.

46. Inca Gold Mining Rate Calculations: Inca gold mining rate calculations can be used to teach rate, time, and historical mathematics.

47. Native American Suni Jewellery Symmetry: Native American Suni jewellery symmetry can be explored to teach symmetry, patterns, and cultural representations.

48. Aztec Feathered Serpent Symbolism: Aztec feathered serpent symbolism can be studied to introduce symbolism, patterns, and cultural representations.

49. Native American Ojibwe Birch Bark Canoe Geometry: Native American Ojibwe birch bark canoe geometry can be used to teach geometry, measurement, and cultural representations.

50. Inuit Whale Hunting Navigation and Trigonometry: Inuit whale hunting navigation and trigonometry can be explored to teach trigonometry, angles, and cultural practices.

51. Native American Mohican Wampum Belt Number Systems: Native American Mohican wampum belt number systems can be studied to introduce number systems, counting, and cultural representations.

52. Aztec Writing System Logograms: Aztec writing system logograms can be used to teach symbolism, communication, and historical representations.

53. Native American Seminole Patchwork Quilt Patterns: Native American Seminole patchwork quilt patterns can be explored to teach geometry, patterns, and cultural representations.

54. Inca Chasquis Relay System Distance and Speed: Inca Chasquis relay system distance and speed calculations can be studied to introduce distance, speed, and historical mathematics.

55. Native American Choctaw Stickball Field Measurements: Native American Choctaw stickball field measurements can be used to teach geometry, measurement, and cultural representations.

56. Aztec Architecture Symmetry and Proportions: Aztec architecture symmetry and proportions can be explored to teach geometry, symmetry, and historical mathematics.

57. Native American Haida Totem Pole Carving Symmetry: Native American Haida totem pole carving symmetry can be studied to introduce symmetry, patterns, and cultural representations.

58. Inuit Seal Hunting Success Rate Probabilities: Inuit seal hunting success rate probabilities can be used to teach probability, statistics, and cultural practices.

59. Native American Pueblo Pottery Design Rotation Symmetry: Native American Pueblo pottery design rotation symmetry can be explored to teach symmetry, patterns, and cultural representations.

Britain

1. Roman Numerals: Roman numerals can be studied to introduce the numeral system and explore conversions between Roman and Arabic numerals.
2. Stonehenge Geometry: Stonehenge geometry can be explored to teach angles, shapes, and historical mathematics.
3. Celtic Knot Patterns: Celtic knot patterns can be studied to introduce symmetry, patterns, and cultural representations.
4. Medieval Tiling Designs: Medieval tiling designs can be used to teach symmetry, shapes, and historical mathematics.
5. British Pound Currency Conversions: British pound currency conversions can be explored to teach currency, conversions, and real-world applications.
6. Tudor Rose Symmetry: Tudor rose symmetry can be studied to introduce symmetry, patterns, and historical representations.
7. Shakespearean Sonnet Rhyme Schemes: Shakespearean sonnet rhyme schemes can be used to teach patterns, sequences, and cultural representations.
8. British Monarchy Succession Patterns: British monarchy succession patterns can be explored to teach patterns, sequences, and historical mathematics.
9. Royal Guard Changing of the Guard Time Intervals: The Royal Guard changing of the guard time intervals can be studied to teach time, intervals, and cultural practices.
10. English Premier League Football Statistics: English Premier League football statistics can be used to teach data analysis, statistics, and real-world applications.
11. British Imperial Units Conversions: British Imperial units conversions can be explored to teach measurement, conversions, and historical mathematics.
12. Scottish Tartan Weaving Patterns: Scottish tartan weaving patterns can be studied to explore geometry, patterns, and cultural representations.
13. Welsh Eisteddfod Poetry Meter and Rhythm: Welsh Eisteddfod poetry meter and rhythm can be used to teach rhythm, patterns, and cultural representations.
14. Magna Carta Proportional Representation: The Magna Carta proportional representation principles can be explored to teach ratios, proportions, and historical mathematics.
15. English Garden Design Symmetry: English garden design symmetry can be studied to introduce symmetry, patterns, and cultural representations.
16. British Royal Navy Flag Semaphore Alphabet: The British Royal Navy flag semaphore alphabet can be used to teach codes, communication, and historical representations.
17. Scottish Bagpipe Music Time Signatures: Scottish bagpipe music time signatures can be explored to teach rhythm, time signatures, and cultural representations.
18. British Monarchy Heraldry Shield Geometry: British monarchy heraldry shield geometry can be studied to introduce geometry, shapes, and historical representations.

19. London Underground Tube Network Distance and Time Calculations: London Underground tube network distance and time calculations can be used to teach distance, time, and real-world applications.

20. British Breakfast Cooking Time Conversions: British breakfast cooking time conversions can be explored to teach measurement, conversions, and culinary mathematics.

21. English Garden Mase Symmetry and Path Lengths: English garden mase symmetry and path lengths can be studied to introduce symmetry, measurements, and cultural representations.

22. Scottish Highland Games Distance Throwing: Scottish Highland Games distance throwing can be used to teach measurement, distance, and cultural representations.

23. British Royal Family Tree Lineage Charts: British royal family tree lineage charts can be explored to teach data representation, patterns, and historical mathematics.

24. English Literature Character Analysis Charts: English literature character analysis charts can be studied to introduce data analysis, charts, and cultural representations.

25. Welsh Choir Harmonies and Musical Intervals: Welsh choir harmonies and musical intervals can be used to teach intervals, harmonies, and cultural representations.

26. British Museum Artifact Time Periods: British Museum artifact time periods can be explored to teach historical timelines, data analysis, and cultural representations.

27. English Premier League Football Transfer Market Analysis: English Premier League football transfer market analysis can be studied to teach data analysis, statistics, and real-world applications.

28. Scottish Kilt Pleating and Measurement: Scottish kilt pleating and measurement can be used to teach geometry, measurement, and cultural representations.

29. British Royal Mail Postal Rates and Weight Calculations: British Royal Mail postal rates and weight calculations can be explored to teach rates, weights, and real-world applications.

30. Shakespearean Play Wordplay and Puns: Shakespearean play wordplay and puns can be studied to introduce language, wordplay, and cultural representations.

31. British Census Data Analysis: British census data analysis can be used to teach data analysis, statistics, and real-world applications.

32. English Country Manor Estate Area Calculations: English country manor estate area calculations can be explored to teach geometry, area, and cultural representations.

33. Scottish Highland Clan Tartan Colours and Patterns: Scottish Highland clan tartan Colours and patterns can be studied to introduce Colours, patterns, and cultural representations.

34. British Royal Air Force Phonetic Alphabet: The British Royal Air Force phonetic alphabet can be used to teach codes, communication, and historical representations.

35. London Olympics Athletics Records and Analysis: London Olympics athletics records and analysis can be explored to teach data analysis, statistics, and real-world applications.

36. English Literature Sonnet Analysis and Interpretation: English literature sonnet analysis and interpretation can be studied to introduce poetry, analysis, and cultural representations.

37. Scottish Shortbread Baking Recipe Ratios: Scottish shortbread baking recipe ratios can be used to teach ratios, fractions, and culinary mathematics.
38. British Royal Navy Ship Design and Geometry: British Royal Navy ship design and geometry can be explored to teach geometry, shapes, and historical representations.
39. English Literature Shakespearean Sonnet Interpretation: English literature Shakespearean sonnet interpretation can be studied to introduce poetry, interpretation, and cultural representations.
40. Welsh Folklore Storytelling Patterns: Welsh folklore storytelling patterns can be used to teach patterns, sequences, and cultural representations.
41. British Rail Train Timetable Interpretation: British Rail train timetable interpretation can be explored to teach time, schedules, and real-world applications.
42. London Landmarks Architectural Symmetry: London landmarks architectural symmetry can be studied to introduce symmetry, patterns, and cultural representations.
43. Scottish Highland Dance Step Patterns: Scottish Highland dance step patterns can be used to teach patterns, sequences, and cultural representations.
44. British Monarchy Coronation Ceremony Rituals: British monarchy coronation ceremony rituals can be explored to teach historical traditions, sequencing, and cultural representations.
45. English Literature Shakespearean Soliloquy Analysis: English literature Shakespearean soliloquy analysis can be studied to introduce drama, analysis, and cultural representations.
46. Scottish Bagpipe Fingerings and Music Notation: Scottish bagpipe fingerings and music notation can be used to teach musical notation, fingerings, and cultural representations.
47. British Royal Mint Coinage Designs and Symmetry: British Royal Mint coinage designs and symmetry can be explored to teach symmetry, patterns, and historical representations.
48. English Literature Poetry Analysis and Interpretation: English literature poetry analysis and interpretation can be studied to introduce poetic devices, interpretation, and cultural representations.
49. Welsh Folk Music Melody and Rhythm: Welsh folk music melody and rhythm can be used to teach melody, rhythm, and cultural representations.
50. British Museum Artifact Measurements and Estimations: British Museum artifact measurements and estimations can be explored to teach measurement, estimations, and cultural representations.
51. London Marathon Race Pace and Distance Calculations: London Marathon race pace and distance calculations can be used to teach rate, distance, and real-world applications.
52. English Country Garden Plant Spacing and Geometry: English country garden plant spacing and geometry can be studied to introduce geometry, measurements, and cultural representations.
53. Scottish Gaelic Language Syntax and Grammar: Scottish Gaelic language syntax and grammar can be used to teach language, syntax, and cultural representations.

54. British Royal Society Scientific Notation and Units: British Royal Society scientific notation and units can be explored to teach scientific notation, units, and historical mathematics.

55. English Literature Character Development and Analysis: English literature character development and analysis can be studied to introduce character analysis, interpretation, and cultural representations.

56. Welsh Dragon Flag Geometry: Welsh dragon flag geometry can be used to teach geometry, shapes, and cultural representations.

57. British Royal Navy Navigation and Trigonometry: British Royal Navy navigation and trigonometry can be explored to teach trigonometry, navigation, and historical mathematics.

58. London Underground Tube Fare Calculations: London Underground tube fare calculations can be studied to introduce rates, calculations, and real-world applications.

59. English Literature Gothic Novel Symbolism: English literature Gothic novel symbolism can be used to teach symbolism, interpretation, and cultural representations.

60. Scottish Ceilidh Dance Patterns: Scottish ceilidh dance patterns can be explored to teach patterns, sequences, and cultural representations.

1. Babylonian Mathematics: Babylonian mathematics can be studied to introduce the base-60 number system and explore ancient methods of calculation.

2. Islamic Geometric Patterns: Islamic geometric patterns can be used to teach symmetry, tessellations, and cultural representations.

3. Arabic Number System: The Arabic number system, including the concept of zero, can be explored to teach place value, operations, and number sense.

4. Al-Khwarizmi's Algebra: Al-Khwarizmi's algebra can be studied to introduce algebraic equations, equations solving techniques, and their historical significance.

5. Astrolabe Navigation: Astrolabe navigation can be used to teach angles, astronomy, and historical mathematics.

6. Persian Carpet Weaving Symmetry: Persian carpet weaving symmetry can be explored to teach symmetry, patterns, and cultural representations.

7. Sine and Cosine Trigonometric Functions: The development of sine and cosine functions by Muslim mathematicians can be studied to teach trigonometry and their applications.

8. Omar Khayyam's Algebraic Equations: Omar Khayyam's algebraic equations can be used to teach quadratic equations, factoring, and historical contributions.

9. Islamic Calligraphy Symmetry: Islamic calligraphy symmetry can be explored to teach symmetry, patterns, and cultural representations.

10. Islamic Calendar Lunar Calculations: Islamic calendar lunar calculations can be studied to introduce calendar systems, time calculations, and cultural practices.

11. Alhazen's Optics and the Law of Reflection: Alhazen's work on optics and the law of reflection can be used to teach geometric optics, the law of reflection, and historical contributions.

12. Arabesque Art Patterns: Arabesque art patterns can be explored to teach symmetry, patterns, and cultural representations.

13. Al-Jabr (Algebra): The term "algebra" derived from the Arabic word "al-jabr" introduced by Muslim mathematicians can be studied to teach algebraic concepts, equations, and their historical roots.

14. Islamic Architecture Geometry: Islamic architecture geometry can be used to teach geometry, shapes, and cultural representations.

15. Muqarnas Geometric Designs: Muqarnas geometric designs can be explored to teach symmetry, geometry, and cultural representations.

16. The Rhind Mathematical Papyrus: The Rhind Mathematical Papyrus, an ancient Egyptian mathematical document, can be studied to introduce fractions, unit conversions, and historical mathematics.

17. Persian Arithmetic Calculation Techniques: Persian arithmetic calculation techniques can be used to teach efficient mental calculations, estimation, and historical methods.

18. Arab Mathematician Al-Kindi's Cryptography: Al-Kindi's cryptography methods can be explored to teach codes, encryption, and historical contributions.

19. Islamic Art Tessellations: Islamic art tessellations can be studied to introduce symmetry, patterns, and cultural representations.

20. Arab Mathematician Ibn Al-Haytham's Optics: Ibn Al-Haytham's work on optics can be used to teach light, vision, and historical contributions.

21. Al-Mahani's Trigonometric Identities: Al-Mahani's trigonometric identities can be explored to teach trigonometric functions, identities, and historical contributions.

22. Islamic Architecture Proportions: Islamic architecture proportions, such as the golden ratio, can be studied to introduce proportions, ratios, and cultural representations.

23. Persian Mathematician Al-Kashi's Approximation Techniques: Al-Kashi's approximation techniques can be used to teach estimation, rounding, and historical methods.

24. Islamic Astronomy: Islamic astronomy can be explored to teach celestial objects, measurement of time, and historical contributions.

25. The Siggurat Construction Geometry: The geometry involved in the construction of siggurats in ancient Mesopotamia can be studied to introduce geometric shapes, measurements, and historical architecture.

26. Islamic Medicine and Algebra: The integration of algebra in Islamic medicine can be used to teach the application of algebraic concepts in real-world contexts.

27. Ottoman Empire Architectural Designs: Ottoman Empire architectural designs can be explored to teach geometry, symmetry, and cultural representations.

28. Arab Mathematician Ibn Al-Banna's Arithmetic: Ibn Al-Banna's arithmetic methods can be studied to introduce arithmetic algorithms, calculations, and historical contributions.
29. Islamic Tiling Patterns: Islamic tiling patterns can be used to teach symmetry, tessellations, and cultural representations.
30. Persian Mathematician Sharaf Al-Din Al-Tusi's Trigonometry: Al-Tusi's trigonometry methods can be explored to teach trigonometric functions, measurements, and historical contributions.
31. Al-Hasib Al-Karaji's Binomial Theorem: Al-Karaji's binomial theorem can be studied to introduce binomial expansion, coefficients, and historical contributions.
32. Islamic Legal System Calculations: Islamic legal system calculations, such as calculating inheritances, can be used to teach percentage, proportions, and real-world applications.
33. Arab Mathematician Al-Biruni's Measurement Techniques: Al-Biruni's measurement techniques can be explored to teach measurement, units, and historical contributions.
34. Islamic Mosaics: Islamic mosaics can be studied to introduce patterns, symmetry, and cultural representations.
35. Al-Sahrawi's Medical Calculations: Al-Sahrawi's medical calculations can be used to teach measurement conversions, dosage calculations, and historical contributions in medicine.
36. Persian Mathematician Al-Zamawal's Algebra: Al-Zamawal's algebraic equations can be explored to teach equations, variables, and historical contributions.
37. Islamic Bookbinding Geometric Designs: Islamic bookbinding geometric designs can be studied to introduce symmetry, patterns, and cultural representations.
38. Arab Mathematician Ibn Sina's Trigonometry: Ibn Sina's trigonometric methods can be used to teach trigonometric functions, measurements, and historical contributions.
39. Al-Khasini's Balance Scale and Archimedes' Principle: Al-Khasini's work on balance scales and Archimedes' principle can be explored to teach measurements, weight, and historical contributions.
40. Islamic Music Maqam Scales: Islamic music maqam scales can be studied to introduce scales, intervals, and cultural representations.
41. Persian Mathematician Al-Biruni's Calculations of Earth's Radius: Al-Biruni's calculations of the Earth's radius can be used to teach measurements, calculations, and historical contributions.
42. Al-Karaji's Factorisation Methods: Al-Karaji's factorisation methods can be explored to teach factorisation, prime numbers, and historical contributions.
43. Islamic Manuscript Illumination Symmetry: Islamic manuscript illumination symmetry can be studied to introduce symmetry, patterns, and cultural representations.
44. Arab Mathematician Ibn Al-Haytham's Camera Obscura: Ibn Al-Haytham's work on the camera obscura can be used to teach optics, geometry, and historical contributions.

45. Persian Architectural Minaret Designs: Persian architectural minaret designs can be explored to teach geometry, proportions, and cultural representations.
46. Al-Hasan Ibn Al-Haytham's Visual Perception Theory: Al-Hasan Ibn Al-Haytham's theory of visual perception can be studied to introduce optics, vision, and historical contributions.
47. Islamic Manuscript Page Layout Geometry: Islamic manuscript page layout geometry can be used to teach geometry, shapes, and cultural representations.
48. Arab Mathematician Thabit ibn Qurra's Number Theory: Thabit ibn Qurra's number theory can be explored to teach prime numbers, divisors, and historical contributions.
49. Persian Architectural Iwan Designs: Persian architectural iwan designs can be studied to introduce symmetry, shapes, and cultural representations.
50. Al-Ma'mun's Astronomical Observations: Al-Ma'mun's astronomical observations can be used to teach celestial objects, measurements, and historical contributions.
51. Islamic Legal Inheritance Calculations: Islamic legal inheritance calculations can be explored to teach algebraic equations, percentages, and real-world applications.

Teachers can select specific traditions, equations, and explanations to align with relevant curriculum topics, engage students in cultural exploration, and make mathematics learning more meaningful and relevant.

Australia/Oceania

1. Indigenous Australian Dot Painting Symmetry: Indigenous Australian dot painting symmetry can be studied to introduce symmetry, patterns, and cultural representations.
2. Polynesian Navigation Techniques: Polynesian navigation techniques can be used to teach angles, trigonometry, and historical mathematics.
3. Aboriginal Dreamtime Stories and Time Calculations: Aboriginal Dreamtime stories and time calculations can be explored to teach time calculations, calendars, and cultural representations.
4. Maori Carving and Geometry: Maori carving and geometry can be studied to introduce geometry, shapes, and cultural representations.
5. Torres Strait Islander Seasonal Calendar: Torres Strait Islander seasonal calendar can be used to teach time, seasons, and cultural representations.
6. Aboriginal Counting Systems: Aboriginal counting systems, such as the Yolngu matha system, can be explored to teach number systems, place value, and cultural mathematics.
7. Maori Whakairo (Wood Carving) Patterns: Maori whakairo patterns can be studied to introduce symmetry, patterns, and cultural representations.

8. Polynesian Canoe Design and Geometry: Polynesian canoe design and geometry can be used to teach geometry, measurements, and historical mathematics.

9. Aboriginal Art and Number Patterns: Aboriginal art and number patterns can be explored to teach patterns, sequences, and cultural representations.

10. Maori Wharepuni (Meeting House) Proportions: Maori wharepuni proportions can be studied to introduce proportions, ratios, and cultural representations.

11. Aboriginal Kinship Systems and Graph Theory: Aboriginal kinship systems and graph theory can be used to teach graph theory, networks, and cultural representations.

12. Polynesian Star Maps and Celestial Navigation: Polynesian star maps and celestial navigation can be explored to teach astronomy, angles, and historical mathematics.

13. Torres Strait Islander Weaving Patterns: Torres Strait Islander weaving patterns can be studied to introduce patterns, symmetry, and cultural representations.

14. Aboriginal Bush Medicine and Measurement Conversions: Aboriginal bush medicine and measurement conversions can be used to teach measurement, conversions, and real-world applications.

15. Maori Whakapapa (Genealogy) and Tree Diagrams: Maori whakapapa and tree diagrams can be explored to teach tree diagrams, relationships, and cultural representations.

16. Polynesian Tattoos and Symmetry: Polynesian tattoos and symmetry can be studied to introduce symmetry, patterns, and cultural representations.

17. Torres Strait Islander Drum Beats and Rhythm: Torres Strait Islander drum beats and rhythm can be used to teach patterns, sequences, and cultural representations.

18. Aboriginal Songlines and Geographical Coordinates: Aboriginal songlines and geographical coordinates can be explored to teach coordinates, mapping, and cultural representations.

19. Maori Whenua (Land) and Geometric Properties: Maori whenua and geometric properties can be studied to introduce geometric properties, shapes, and cultural representations.

20. Polynesian Tapacloth Patterns: Polynesian tapacloth patterns can be used to teach symmetry, patterns, and cultural representations.

21. Torres Strait Islander Traditional Fishing Techniques and Measurement: Torres Strait Islander traditional fishing techniques and measurement can be explored to teach measurement, estimation, and cultural practices.

22. Aboriginal Firestick Farming and Area Calculations: Aboriginal firestick farming and area calculations can be studied to introduce area calculations, land management, and cultural practices.

23. Maori Whakairo (Carving) Symmetry: Maori whakairo symmetry can be used to teach symmetry, patterns, and cultural representations.

24. Polynesian Lei Making and Perimeter: Polynesian lei making and perimeter can be explored to teach perimeter, measurements, and cultural representations.

25. Torres Strait Islander Mask Designs: Torres Strait Islander mask designs can be studied to introduce shapes, patterns, and cultural representations.

26. Aboriginal Rock Art and Geometric Shapes: Aboriginal rock art and geometric shapes can be used to teach geometric shapes, patterns, and cultural representations.

27. Maori Whakapapa (Genealogy) and Probability: Maori whakapapa and probability can be explored to teach probability, tree diagrams, and cultural representations.

28. Polynesian Dance Rhythms and Fractions: Polynesian dance rhythms and fractions can be studied to introduce fractions, patterns, and cultural representations.

29. Torres Strait Islander Drumming Patterns: Torres Strait Islander drumming patterns can be used to teach patterns, sequences, and cultural representations.

30. Aboriginal Bush Tucker and Data Handling: Aboriginal bush tucker and data handling can be explored to teach data collection, analysis, and cultural practices.

31. Maori Wharepuni (Meeting House) Construction Geometry: Maori wharepuni construction geometry can be studied to introduce geometric shapes, measurements, and cultural representations.

32. Polynesian Tapa Cloth Measurements: Polynesian tapa cloth measurements can be used to teach measurements, conversions, and cultural representations.

33. Torres Strait Islander Games and Probability: Torres Strait Islander games and probability can be explored to teach probability, outcomes, and cultural representations.

34. Aboriginal Rock Art and Time Periods: Aboriginal rock art and time periods can be studied to introduce timelines, historical periods, and cultural representations.

35. Maori Poi Dancing and Angles: Maori poi dancing and angles can be used to teach angles, geometry, and cultural representations.

36. Polynesian Weaving and Symmetry: Polynesian weaving and symmetry can be explored to teach symmetry, patterns, and cultural representations.

37. Torres Strait Islander Navigation Charts and Mapping: Torres Strait Islander navigation charts and mapping can be studied to introduce mapping, coordinates, and cultural representations.

38. Aboriginal Seasonal Fire Management and Data Analysis: Aboriginal seasonal fire management and data analysis can be used to teach data analysis, trends, and cultural practices.

39. Maori Ta Moko (Tattoo) Patterns: Maori ta moko patterns can be explored to teach symmetry, patterns, and cultural representations.

40. Polynesian Drumming Patterns and Sequences: Polynesian drumming patterns and sequences can be studied to introduce patterns, sequences, and cultural representations.

41. Torres Strait Islander Art and Shape Transformations: Torres Strait Islander art and shape transformations can be used to teach transformations, shapes, and cultural representations.

42. Aboriginal Rock Art and Graphs: Aboriginal rock art and graphs can be explored to teach graphing, data representation, and cultural representations.

43. Maori Wharepuni (Meeting House) Proportional Reasoning: Maori wharepuni proportional reasoning can be studied to introduce proportional relationships, scaling, and cultural representations.

44. Polynesian Tiki Carving Symmetry: Polynesian tiki carving symmetry can be used to teach symmetry, patterns, and cultural representations.

45. Torres Strait Islander Basket Weaving and Measurement: Torres Strait Islander basket weaving and measurement can be explored to teach measurement, conversions, and cultural practices.

46. Aboriginal Body Painting Patterns: Aboriginal body painting patterns can be studied to introduce patterns, symmetry, and cultural representations.

47. Maori Whakapapa (Genealogy) and Venn Diagrams: Maori whakapapa and Venn diagrams can be used to teach set theory, relationships, and cultural representations

Aboriginal and Torres Strait Islander Mathematics:

1. Introduce students to the mathematical knowledge and practices of Indigenous Australians, such as their unique counting systems or their geometric patterns. Discuss the significance of these mathematical traditions in their culture.

2. Pacific Island Mathematics: Explore the traditional navigation techniques used by Pacific Island communities, such as the star compass or stick charts. Explain how these techniques rely on mathematical principles and demonstrate their practical applications.

3. Maori Mathematics (New Zealand): Teach students about traditional Maori games involving counting, probability, and strategy. For example, introduce games like Mancala or variations of Morris games, which can enhance students' understanding of number patterns and strategic thinking.

4. Papuan Mathematical Systems: Introduce students to the complex number systems used by various Papua New Guinea communities. Discuss their unique ways of representing and manipulating numbers, which can offer alternative perspectives on mathematical concepts.

5. Mathematics in Polynesian Weaving: Explore the intricate patterns and designs found in Polynesian weaving. Discuss the underlying mathematical principles, such as symmetry, shapes, and proportions, and how they can be related to geometric concepts in the GCSE curriculum.

6. Mathematical Representations in Aboriginal Art: Analyse the mathematical concepts reflected in Aboriginal art, such as dot paintings or sand drawings. Explore concepts like symmetry, tessellations, or spatial reasoning, and relate them to relevant topics in the GCSE Mathematics curriculum.

Annotated Bibliography: Decolonising Mathematics Education in the UK

1. **Book: "Decolonising the Mind" by Ngũgĩ wa Thiong'o**
    - This seminal work explores the effects of colonialism on language, literature, and education in Africa. Ngũgĩ argues for the decolonisation of education as a means of reclaiming cultural identity and empowering marginalised communities. While not focused specifically on mathematics education, the principles and perspectives discussed are highly relevant to broader efforts to decolonise education.

2. **Article: "Decolonising the Mathematics Curriculum" by Rachel Bateman and Nuala Burgess**
    - This article examines the implications of colonialism for mathematics education and offers strategies for decolonising the curriculum. Bateman and Burgess discuss the importance of incorporating diverse cultural perspectives, challenging Eurocentric narratives, and promoting critical consciousness among students. They provide practical examples and case studies of decolonising initiatives in mathematics classrooms.

3. **Research Paper: "Decolonising Mathematics Education: Addressing Eurocentrism in Mathematics" by Shirin Richter**
    - Richter's research paper critically examines the Eurocentric biases embedded in mathematics education and explores ways to decolonise the curriculum. She argues for a more inclusive approach that acknowledges the contributions of non-Western cultures to mathematics and challenges the hegemony of Western mathematical knowledge. The paper offers theoretical insights and practical recommendations for educators.

4. **Book: "Culturally Responsive Mathematics Education" edited by Brian Greer**
    - This edited volume brings together scholars and practitioners to explore the intersection of culture and mathematics education. The contributors discuss strategies for making mathematics more culturally responsive, including incorporating students' cultural backgrounds into the curriculum, promoting critical consciousness, and addressing issues of power and privilege in the classroom. The book provides theoretical frameworks and practical examples for educators seeking to decolonise mathematics education.

5. **Article: "Decolonising Mathematics: Experiences from India" by Renuka Vithal and Renuka Sane**
    - Drawing on their experiences in India, Vithal and Sane discuss the challenges and opportunities of decolonising mathematics education in a post-colonial context. They explore the role of language, culture, and history in shaping mathematical

knowledge and argue for a more inclusive approach that acknowledges multiple ways of knowing and learning mathematics. The article offers insights into the complexities of decolonising mathematics education in diverse cultural contexts.

Organisations and Initiatives in the UK Focused on Decolonising Education

1. **Decolonising Education Network (DEN)**
   - DEN is a grassroots organisation dedicated to promoting decolonisation in education across the UK. They organise workshops, seminars, and events focused on challenging colonial narratives, promoting cultural diversity, and empowering marginalised communities within educational institutions.

2. **Decolonise the Curriculum**
   - Decolonise the Curriculum is a student-led initiative advocating for the decolonisation of university curricula in the UK. They campaign for greater representation of non-Western perspectives and histories in course content and work with academic departments to implement decolonising initiatives.

3. **Runnymede Trust**
   - The Runnymede Trust is a leading UK think tank focused on race equality and social justice. They conduct research, produce reports, and engage in advocacy to promote understanding and challenge racial inequalities in education, including efforts to decolonise the curriculum and diversify educational leadership.

4. **Black Educators Alliance (BEA)**
   - BEA is a network of Black educators working to address racial disparities and promote inclusivity in education. They provide professional development, networking opportunities, and resources for educators interested in decolonising the curriculum and creating more culturally responsive learning environments.

5. **The Free Black University**
   - The Free Black University is a community-led initiative that offers free courses, workshops, and events focused on Black history, culture, and liberation. They provide resources and support for educators seeking to incorporate Black perspectives and experiences into their teaching practice and curriculum design.